Introduction to Pollen Spore Research

Introduction to Pollen Spore Research

Dushyant Kumar Saxena

CYBER TECH PUBLICATIONS
4264/3, Ansari Road, Daryaganj, New Delhi-110002 (India)
Ph.: 011-23244078. Fax: 011-23280028
E-Mail: cyberpublicationsdelhi@yahoo.com

INTRODUCTION TO POLLEN SPORE RESEARCH

Dushyant Kumar Saxena

First Edition 2009

Published by :

G.S. Rawat for **Cyber Tech Publications**
4264/3, Ansari Road, Daryaganj, New Delhi-110002 (India)
Ph.: 011-23244078. Fax: 011-23280028
E-Mail: cyberpublicationsdelhi@yahoo.com

Printed at : **Amit Enterprises**, Maujpur, Delhi, (9810878184)

Preface

The family cucurbitaceae is unique in many morphological characters and has received considerable attention at the hands of Indian workers. In India its members occur in wild form and are also widely cultivated for their fruits which are a rich source of vegetables and dessert. Palynologically the family is quite interesting. The characters of pollen morphology have proved useful in identification and in solving taxonomic problems. These have been used effectively in the classification of family in recent years. Physiologically also cucurbit pollen has remarkable features of a polysiphouous germination, branching of pollen tube and its persistence. There are several studies on pollen germination, pollen tube growth and pollen storage. All this information occurs in research publications in different journals.

"The work that forms the basis of the present publication was carried out at the by a research team under the strict supervision. A sincere effort was made to obtain samples of pollen material of wild cucurbits, from localities throughout the country. Samples were collected for large number of cultivars of cultivated species of Cucumis, Citrullus etc. Various aspects of male structure including anther development, pollen morphology and pollen physiology were investigated. The author made a thorough survey of literature and has really come up with a good compilation of data based on his own findings and that of other researchers, particularly the Indian work.

—*Author*

CONTENTS

Preface

1. Global Observations of Forest Cover 1
2. Fucus Vesiculosus in the Atidal Baltic Sea 42
3. Status and Culture of Tropical Agarophytes 54
4. Status and Culture of Agarophytes in China 67
5. Status and Culture of Agarophytes in Indonesia 76
6. Small-scale Agarophyte Processing 93
7. Extraction of Agar from Gracilaria Edulis as a Village Level Technology 110
8. Management of Natural Resources of Tropical Agarophytes 138
9. Germplasm Collection of Capsicum 153
10. Pollen Cryopreservation in Capsicum Species 185
11. Effects of Cryopreservation of Pepper and Eggplant Seeds 196
12. Integrating Science and Management of Biodiversity in the Western Ghats 220
13. Scelionidae of India-An Overview 238
14. Impacts of Forest Cover Loss on Hydrologic Processes 255
15. Pollinators: Neglected Biodiversity of Importance to Food and Agriculture 275

Bibliography 289

Index 291

1

Global Observations of Forest Cover

INTRODUCTION

Two of the most crucial roles of the Committee on Earth Observation Satellites (CEOS) are to promote the use of Earth observation data to improve the quality of life on earth, and to increase coordination and reduce duplication of the satellite programs of the member states. At its 1995 plenary meeting in Montreal, Canada, CEOS recognized the need to establish a strategic framework to determine and assess the data and information requirements for satellite-based Earth observations, and to facilitate the transition to routine operation of those research systems which have proven most useful.

The plenary also recognized the need to improve the coordination of the space-based and in situ components of Earth observations, and to increase the terrestrial benefits of the Earth-observation data already being collected (Shaffer, 1996).

In response to the requirements identified at the Montreal plenary, a working group to discuss an Integrated Global Observing Strategy (IGOS) was convened in March, 1996.

This group recognized that the most effective means to accomplish its mandated objectives would be to identify pilot projects dealing with key areas of earth observations, and recommended the establishment of a Strategic Implementation Team (SIT) to select pilot projects and an Analysis Group (AG) to coordinate the requirements of the pilot projects and to ensure good coordination between producers and users of earth-observation data (Shaffer, 1997). The general goals of the pilot projects have been identified by the SIT:

(1) Provide a framework for a coherent set of user requirements so that providers can respond to them.

(2) Reduce unnecessary duplication of observations.

(3) Assist in the improved allocation of resources among different types of observation systems.

(4) Make possible the creation of improved higher level products by facilitating the integration of multiple data sets from different agencies, national, and international organizations.

(5) Provide a framework for decisions on continuity and spatial comprehensiveness of key observations.

(6) Identify situations where existing international arrangements do not exist for the management and distribution of key global observations and products.

(7) Assist in the transitioning of systems from research to operational status through improved international cooperation.

(8) Provide improved understanding for Governments on the need for global observation through the presentation of an overarching view of their capabilities and limitations.

SIT held its first meeting in Irvine, California, in February, 1997. The attendees identified six pilot projects1 of vital importance, whose execution is expected to advance the CEOS objectives stated above. The Canadian Space Agency (CSA) and the Canada Centre for Remote Sensing (CCRS) agreed to co-lead a pilot project on Global Observations of Forest Cover (GOFC), with the U.S. National Aeronautics and Space Administration (NASA) agreeing to play a strong supporting role (Guertin, 1997).

Recognizing its potential importance to a wide spectrum of users, this program was subscribed to by numerous space agencies including CSA/CCRS, NASA, ESA, INPE, CNES,

EUMETSAT, NASDA, and ASI. It has also been endorsed by representatives of GTOS, GCOS, TOPC, EC, and FAO. However, a program of global observations of forest cover is also recognized as very challenging, requiring coordination of acquisition and analysis of data from satellites from many nations operating in the visible, infrared, and microwave regions of the spectrum. It also requires considerable input of in situ data and local knowledge of land cover characteristics. Ultimately, the GOFC is envisioned to lead to a routine operational capability for the observation of changes in global forest cover and its attributes.

OBJECTIVES OF THE PILOT PROJECT ON GLOBAL OBSERVATIONS OF FOREST COVER

As a contribution to the implementation of an ongoing global forest monitoring system that will meet specific needs of international conventions, global observing systems, national institutions, and supporting research, GOFC will

(1) Produce high quality, multi-resolution, multi-temporal global data sets and derived products of forest cover and attributes;

(2) With particular attention to areas of rapid change2 and fragmentation;

(3) To be repeated for quantitative analysis of variation on a three to ten year cycle;

(4) With associated regional applications and methodological investigations;

(5) For the benefit of multiple user communities;

(6) Ultimate transition to routine operational use.

These six objectives will enable the Pilot Project to conduct an end-to-end test of many aspects of the spacebased observational systems (including data systems) that will be required for an operational IGOS. It will also enable the CEOS partners to evaluate how their space-based observations can effectively be coupled to in situ data, and how data products can be provided effectively to several important user communities.

SCIENTIFIC, POLICY, AND RESOURCE MANAGEMENT NEEDS

The potential user community is extremely diverse. In many ways, the GOFC must assist communities of users in extracting the maximum information available from remotely sensed data, and from its combination with in situ data. Likewise, the various communities of users themselves can provide essential guidance to the GOFC to focus on those products and data sets that are of the most benefit to them, in order to guide the efficient allocation of resources, and prepare most effectively for the transition to operational status. The ultimate success of this program will lie in the extent to which national, sub-national, and international organizations make routine, ongoing use of earth-observation data to satisfy their information needs. It will be critical in both respects to build on those programs that already exist for the production and analysis of data on forest cover and attributes. There is an explicit desire to learn from the vast experience available in the remote sensing, in situ, research, and operational communities.

Issues

(1) Global Change. There are three pathways of relevance for global change and forest lands with implications for this CEOS project (Turner et al., 1995). The first is the interaction of forests and the atmosphere: regulation of the hydrologic cycle and energy budget, with implications for weather and climate prediction. Understanding these interactions is a crucial part of the global change science agenda. The second is both a scientific and policy issue: the implications of changes in forest land for the atmospheric carbon dioxide budget. Are forested lands net sources of carbon or net sinks of carbon? What are the rates of deforestation and reforestation? What are the implications of changes in fire frequency? The third pathway is the potential impacts of climate change: what effect will changes in temperature, precipitation, and concentrations of carbon dioxide and other radiatively active gases have on forest composition,

productivity, health, and distribution, and therefore on the economic activities that are generated by forested lands? What feedbacks might these changes have to the climate system itself?

(2) Timber, fuel, and fibre. For centuries, human society has depended critically on the continued supply of wood and fiber from forest land. Many societies have declined after depleting their forest resources (Perlin, 1991). The twentieth century has produced substantial improvements in the efficiency of production of wood and fiber from forests. Increasing populations and increasing literacy and affluence in the twenty-first century will continue to put increasing demands on the forests of the world. Continued improvements in forest management will be necessary to meet these demands without depleting the forest resource (Nyland, 1996).

(3) Watershed Protection. Forested areas and other wooded lands play a crucial role for terrestrial hydrological systems (Nyland, 1996; Hammond 1991). These areas constitute the most important buffer zones for both runoff and infiltration to ground water reserves. The actual location and size of forests in relation to the watershed often determines the run-off fraction of the precipitation and becomes the dominant factor in flood and erosion prevention. A close monitoring of forests in relation to the hydrographic network thus becomes a critical issue in physical planning, landscape planning, and environmental protection in all regions of the world.

(4) Biodiversity. Concern over the rapid, human-driven loss of biodiversity worldwide is rapidly mounting, culminating in the International Convention on Biological Diversity (Interim Secretariat of the Convention on Biological Diversity, 1994). The extent and condition of forested lands are central issues to the preservation and sustainable use of biodiversity. Both forest loss and forest fragmentation have been implicated in losses of biodiversity by the scientific community. On the other hand, many national programs for sustainable

forest management are now incorporating biodiversity as one of the attributes to be enhanced and maintained. (e.g. Canadian Council of Forest Ministers, 1995.)

(5) Recreation and Tourism. Forests provide habitat for wildlife, and opportunities for recreation and tourism. In many countries, the direct economic return from these activities approaches the return from extraction of wood and fibre. Many other countries want to increase the benefits they derive from recreation and tourism. Information obtained from Earth observation data can aid in the strategic planning for recreation and tourism in forested areas.

(6) Sustainable Forest Management. Sustainable forest management has become an extremely important philosophy, along with ecosystem management, to guide the plans of land-managers (both public and private) for the use of forested lands over the next several decades. Sustainable forest management recognizes the multiple benefits provided by forests, and strives to respond to the needs and wants of the multiple stakeholders who benefit from the forest, while conserving its resources so future generations can also benefit from the forest.

Earth observation data, in conjunction with geographic information systems, provides exciting new opportunities to engage a broad spectrum of stakeholders in long-term forest management planning.

User Communities

The GOFC planning group has identified several policy-level issues and potential user communities that together provide the core set of requirements for data sets, derived products, and subsequent analyses. Table 1 illustrates those user communities that have clearly articulated their observational requirements for forest information.

(1) *International Environmental Conventions and Agreements.* The major conventions concerned with global forests are the Convention on Biological Diversity

(Interim Secretariat for the Convention on Biological Diversity), on Climate Change, and on Desertification. Their core information needs are those quantifying major changes in the distribution and function of forests over time. Some of these can be met with Earth observation data, and are in general consistent with those discussed below for the GxOS's. The Framework Convention for Climate Change, particularly the Kyoto Protocol, (Framework Convention for Climate Change Internet Website), the Biological Diversity Convention (Biological Diversity Convention Internet Website), the Forest Principles agreed to at UNCED, the Montreal, Helsinki, and Tarapoto processes (Canadian Forest Service, 1996; ISCI, 1996; World Forestry Congress, 1997), and others all have a shared need for information on global forests. This need is most explicit in the scientific assessments that are either part of the Conventions and other instruments themselves, as in IPCC, or have been commissioned independently of them, as in the Global Biodiversity Assessment. The specific national requirements of the FCCC for the development and monitoring of national emissions inventories, and the international discussions of carbon trading, will require accurate and replicable assessments of forest cover at both national and international levels. The reporting requirements of the Biodiversity Convention for assessing national biodiversity and habitat, likewise require accurate information on forest cover, fragmentation, and change. The Intergovernmental Forum on Forests meets to discuss an Internaltional Convention on Forests. It has emphasized the need for a correct assessment of forest resources.

(2) *International Policy and Operational Institutions.* The Food and Agriculture Organization of the United Nations (UN/FAO) is mandated to produce a global forest resource assessment (FRA) every ten years. Remote sensing observations were used for the tropical regions in 1990 and will be expanded to temperate and boreal

forests in 2000. There is an obvious commonality of interest in access to data and land cover classifications between FAO and GOFC. FAO also carries out projects of capacity building for forest assessment at the national level in developing countries. The provision of standardized regional datasets could facilitate FAO activities at the national level. The Global Observing Systems incorporate scientific, operational, and policy-level concerns at both international and national levels. They have already indicated their strong need for global forest cover information and information on forest attributes, primarily through the report of the Terrestrial Observations Panel for Climate (TOPC), a joint GCOS-GTOS activity (Terrestrial Observation Panel for Climate, 1997). The Global Terrestrial Observing System (GTOS) and the Global Climate Observing System (GCOS) are the major clients in this area. GTOS is an international organization co-sponsored by UNEP, UNESCO, FAO, ICSU, and WMO. GTOS promotes improved collection of and access to bio-physical and socio-economic georeferenced data, with an initial emphasis on existing research centres (Global Terrestrial Observing System, 1997). GTOS requires detailed land cover and land cover change information for many purposes, including the assessment of land use and its impact on the natural ecosystems. Information on forest harvest is also an important input in this process. GCOS is a partnership of WMO, IOC of UNESCO, UNEP and ICSU. The objectives of the Global Climate Observing System, GCOS, are to insure the acquisition of data for climate system monitoring, climate change detection and response monitoring, application of climate information to national economic development, and research. (Global Climate Observing System Internet Website, 1997). GCOS requires information on land cover (coarse resolution) for global circulation modeling and model validation; for carbon, water, and energy modeling; and for the assessment of climate impact. Both GCOS and GTOS require information on net primary productivity as a fundamental parameter linking the terrestrial and the atmospheric systems on the one hand, and the natural and economic systems on the other. Both also require knowledge of forest biomass and fires as major parameters that quantify the distribution of the carbon stocks and the effect of major disturbance agents. The United Nations Environment Programme (UNEP) provides access to datasets and carries out projects in key areas of environmental concern. The satellite data and many of the derived products proposed by GOFC will be be of interest to UNEP.

(3) *Scientific Community.* The International Geosphere-Biosphere Programme (IGBP) and World Climate Research Programme (WCRP) have each articulated clearly the urgent need for comprehensive, spatially explicit, global data sets on forest cover and attributes. Several projects in the International Geosphere-Biosphere Program (IGBP) and the World Climate Research Program (WCRP) require information on global forests. The IGBP Land Use and Cover Change (LUCC) project has critical needs for high-resolution information on land cover and change (Turner et al., 1995). Information on biomass, fires, net primary productivity and land cover change is essential for the IGBP International Global Atmospheric Chemistry (IGAC) project (IGBP, 1994). The IGBP project on Global Change and Terrestrial Ecosystems has strong requirements for land cover, land cover change, leaf area index (LAI), and net primary productivity (NPP) data sets. The Biospheric Aspects of the Hydrological Cycle (BAHC) project has essential needs for land cover, biomass, LAI and NPP data sets (IGBP, 1993). The Global Energy and Water Cycle Experiment (GEWEX) has essential requirements for several forest data sets including land cover and cover change, LAI, biomass, and fires. Several Framework Activities (IGBP Global Analysis and Integrative Modeling, IGBP-Data and Information

System, ISLSCP Initiative II) have especially strong scientific requirements for such forestry data. Furthermore, numerous regional programs such as the Amazon LBA, the European FIRS, and others have essential requirements for most of the above mentioned data sets. Individual national scientific programs, such as the US Global Change Research Program, and the Canadian Global Change Program share these requirements, and place a high priority on the production of such data sets.

(4) *National and Sub-National Institutions.* Countries are interested in their forests because of the value of the forests for their economies and because of the environmental values of forests, both nationally and internationally. Some countries are increasingly concerned about the impact of the environmental movement on the economic utilization of the forest resource. Following the UNCED conference in Rio de Janeiro in 1992, forest nations have been discussing draft forest principles, and developing criteria and indicators of sustainable forest development. For these and other reasons, nations have begun to pursue initiatives aimed at tracking and reporting on the forest condition and changes in their countries. Land cover, cover change and biomass are fundamental 'state'attributes for this purpose. Forest conversion to agriculture, harvest and fires are the most ubiquitous change agents at the global level. Damage by insects and disease, and various forms of pollution are often important nationally and locally, and in several instances are severe enough that affected areas can be mapped using spaceborne data. In many countries forestland management falls within the jurisdiction of subnational (e.g. state and provincial) agencies. Earth-observation data have been extensively used to provide general information at the national and sub-national level. This use is expected to increase as the cost of the data and related technology decreases, and with increasing

pressure on national and sub-national governments to provide more information, more rapidly, with decreasing budgets.

It should be noted that the degree to which a global product meets the needs of individual countries will vary depending on the information sought. For example, the global products for fires, biomass, net primary productivity and land cover at coarse resolution are more likely to meet the needs of individual countries than those for detailed land cover or land cover change. This distinction suggests various possible mechanisms for collaboration between the global project and individual countries, including a contribution to product development and validation, the development of new, more detailed or tailored products to meet the specific national needs, and technology transfer and capacity building. Many national-level forestry ministries and environmental ministries are in the midst of planning efforts to transition from management of forested lands with the primary goal of timber production to sustainable management of forest resources, combining production of forest products with watershed protection, tourism, recreation, and other non-exploitative uses. In addition, the proliferation of international agreements with implications for forests has created new pressures on national institutions to respond with accurate and timely information. Regionally, countries whose forests contribute substantially to the national economy are cooperating to develop criteria and indicators of sustainable forest management.

Satisfying Requirements

It is important to note that while the CEOS pilot project perceives that these issues and user communities will find the products to be important, it cannot expect to satisfy all the demands of all potential users. Instead, there is a core set of requirements that can be substantially fulfilled by the products we have outlined. Other requirements can be addressed, but may not be completely fulfilled by the first set of planned products. However, we leave open the option of further activities in a later phase to define new sets of products that might satisfy additional requirements. All of the users identified above are generally non-commercial scientific and educational organizations, or governmental or intergovernmental organizations dedicated to the public good. We also clearly anticipate that there may be commercial interests in the Pilot Project outcomes as well. Therefore, it is important to emphasize the expectation that derived products from the Pilot Project will be widely accessible and available without discrimination.

It is also clear that many users may wish to derive their own, tailored products from the underlying data used in the Pilot Project. The existing CEOS principles for non-discriminatory access to data should enable all users to search well-organized Pilot Project archives. The Pilot Project will seek to work with data providers to ensure that all data used in the project are organized in such a fashion that they can be accessed and distributed according to the national policies that govern their use.

THE CURRENT EXPERIENCE BASE

Over the past ten years, the remote sensing community has developed an impressive array of projects dealing with global, regional, and local forest-related issues. Projects have been implemented in North America, Europe, Asia, and South America that have examined such issues as tropical deforestation, forest cover mapping, forest health and vigour, and the extent of inundated forest. As a result, there is a broad spectrum of lessons to be learned for the planning and implementation of the Pilot Project.

Existing Programs and Data Sets

The approach adopted is to build on existing national and international programs and activities. Around the world, experts have been building the scientific, technical, and procedural underpinnings for a forest cover change monitoring system through joint space agency initiatives. The Ottawa workshop recognized that some of the components of the program already exist. Rather than duplicate such efforts, wherever possible, the approach would be to integrate these existing studies into a broader programmatic framework.

World Forest Watch: The World Forest Watch Meeting held in São José dos Campos, Brazil (June 1992) provided an initial high-level international forum for the assessment of current approaches to satellite-based monitoring. This meeting also served as a basis for forwarding recommendations from the technical and scientific communities to the policy makers and government leaders at UNCED. A variety of international participants were represented at the World Forest Watch Conference. The conference concluded that significant technical and methodological advances have been made in recent years, and they are now sufficient to proceed with an observation system which could satisfy both scientific and national-level forest management requirements. One important outcome was the Panamazonia Project, in which Brazil shared its proven technology for mapping deforestation in the Amazon region with its neighbors who share the Amazon Basin. These countries were subsequently able to produce deforestation estimates (Martini, 19??).

Landsat Pathfinder: NASA, in conjunction with the EPA and USGS, began a prototype procedure for using large amounts of high resolution satellite imagery to map the rate of tropical deforestation, one of the most important land cover changes. This activity, called the Landsat Pathfinder Project (Landsat Pathfinder Project Internet Website), builds on experience gained during a proof-of-concept exercise as part of NASA's contribution to the International Space Year/World Forest Watch Project. It focused initially on the Brazilian Amazon, and has been expanded as part of NASA's Earth Observing System activities to cover other regions of the humid tropical forests.

This project has succeeded in demonstrating how to develop wall-to-wall maps of forest conversion and regrowth. The project is now in the process of extending its initial proof-of-concept to a large-area experiment across Central Africa, Southeast Asia and the entire Amazon Basin. The project is acquiring several thousand Landsat scenes at three points in time -- mid 1970s, mid 1980s, and mid 1990s -- to compile a comprehensive inventory of deforestation and secondary growth to support global carbon cycle models. Methodology and procedures have been identified. Although this exercise is being implemented for most of the tropics, it is not an operational global program. In principle, it provides an initial large-scale prototype of an operation program.

TREES: The TRopical Ecosystem Environment Observations by Satellites (TREES) project is currently being implemented as a demonstration of the feasibility of applying space observation techniques to monitoring of land cover and biomass burning (Malingreau et al., 1993).

This project, being sponsored by the European Commission and executed by the Space Applications Institute of the Joint Research Centre (Ispra, Italy), utilizes global coverage with a wide range of sensors including AVHRR, ERS-SAR, JERS-1 SAR, ATSR, Resurs, Landsat, and SPOT. It also focuses on the use of thermal sensors for the detection of fires, and incorporates other indicators of deforestation. The project has developed methods of data assimilation leading to an optimization of the multiple data sources. It has also developed a central repository for data on tropical forests from earth observation sensors and other sources.

The Landsat Pathfinder Project and the TREES Project demonstrate the ready feasibility of developing a global land cover monitoring system for the tropical forests. Coverage of tropical forests must be a paramount objective of any program focused on obtaining improved estimates of emissions of carbon dioxide since 90% of the current biogenic emissions come from tropical forest regions.

IGBP 1 km Land Cover Mapping Project: In conjunction with NOAA and the USGS, NASA has been supporting an IGBP project to acquire global, daily coverage with the AVHRR sensor at 1 km resolution (Belward, 1996). This project is the first of its kind to acquire global, daily coverage. The resultant dataset is being processed at the EROS Data Center and provides the basis for mapping land cover, vegetation phenology, and fires. As part of a land-cover mapping exercise, this project will also develop a multi-level set of land cover maps ranging in spatial resolution from 1 degree down to 2 km.

IGBP High Resolution Data Set: The IGBP has also initiated a project to make high resolution data available to the global change research community through cooperation with CEOS. This project is being initiated as a pilot study to make available each year several hundred individual scenes from SPOT, Landsat, ERS-1, MOS-1, JERS-OPS, and IRS. The project is also preparing a centralized archive system in which users can mount an inquiry for data from all of the aforementioned platforms from one point.

Participation in international programs has been found to be a successful means for implementing international research and collaboration. Participation by scientists in the IGBP, WCRP and IHDP has proven to be effective in mobilizing the international research and monitoring community. The new joint core project of the IGBP and IHDP on Land Use and Land Cover Change (LUCC) could greatly benefit the proposed program (Turner et al., 1995). Similarly, the IGBP Global Land Cover Program is a good example of potential international collaboration geared towards the generation of global land cover data. The NASA/ESA/USGS/IGBP Global 1km AVHRR data base is a good example of internationally coordinated data acquisition.

FAO Forest Resource Assessment: The FAO Forest Resource Assessment for the 1990 period (FRA-90) used a combination of earth observation data and national statistics to derive statistics about forest cover for the 1990 datum, and forest cover change for the period 1980 to 1990 (FAO, 1995). For tropical forests, FAO developed a technique called the Interdependent Interpretation Procedure, which relied primarily on a two-date interpretation of large scale photographic prints of full Landsat scenes. Using training centres in Latin America, Africa, and Asia, FAO was able to train local interpreters and obtain consistent results. The resulting change classes also allowed FAO to estimate transitions between several woody biomass categories for Africa, Latin America, and Asia (FAO, 1996). The FAO Forest Resource Assessment Project provides an important interface to the national forest monitoring communities. Such projects can benefit from the proposed GOFC program as well as help the program.

GAP Analysis Project: The GAP Analysis Project (Scott, Tear, and Davis, eds, 1996) is a land cover mapping project being carried by several consortia of U. S. federal and state agencies, led by the U.S. Geological Survey. The purpose is to infer wildlife habitat from land cover, and identify habitats in not adequately represented in conservation lands. The project has obtained a large amount of valuable experience in catalyzing inter-institutional cooperation, and in processing and classifying large volumes of Landsat data according to varying standards developed by user agencies.

Multi-resolution Land Characterization: A good example of a regional /continental scale land cover mapping study is the recently completed USFS /USGS US Forest Cover Map. This map was produced using the 1km AVHRR data set generated by Loveland et al. (1991). A similar effort is being undertaken in Mexico with the help of in-country collaborators. An Interagency Multi-Resolution Land Characterization (MRLC) project, implemented by the USGS on behalf of a number of federal agencies, is generating land cover data sets for the conterminous US. This data set includes products from coarse and high spatial resolution satellite sensors. This project also provides complete coverage of the United States with TM data.

North American Landscape Characterization Project: Datasets from Landsat, SPOT, AVHRR and other sensors with global and national coverage have created opportunities to exploit data in ways not possible just three or four years ago. For example, the North American Landscape Characterization (NALC) Project (North American Landscape Characterization Project Internet Website) is acquiring and processing complete coverage of the continental United States and Central America with Landsat MSS data at three points in time from the 1970s to the present.

The Global Rainforest Mapping Project: Within the framework of NASDA's Global Forest Mapping Program, the

Global Rain Forest Mapping project is a collaboration between NASDA, ASA/JPL/ASF and the Joint Research Centre (SAI/MTV), with the aim of generating JERS-1 L-band SAR data sets of the global tropical forests.

A complete SAR coverage over the tropical belt - in total some 13,000 scenes - was acquired in 1995-96, including multi-temporal coverages over the Amazon and Congo river basins to capture season effects and to map inundation in flooded forests. SAR mosaics at 100 m resolution are being generated to cover the tropical region in all three continents, for distribution free of charge for research and educational purposes on CD-ROM and over the Internet. The mosaics are radiometrically and geometrically corrected (however not for topographical effects). Land cover classifications are also foreseen. A research program is being coordinated by NASDA for multi-temporal studies at local to continental scales.

Starting in 1997, the Global Boreal Forest Mapping Project (GBFM) forms a follow-on project to the GRFM. The GBFM project aims at covering the boreal forests in Siberia, Northern Europe and North America with JERS-1 SAR. Data acquisitions over the each region of interest have been completed once during the spring and summer of 1997. SAR mosaics at 100 m resolution will be generated and distributed free of charge for research and education. Also the GBFM project is a joint effort by NASDA, NASA/JPL/ASF and the JRC, with collaboration from DLR, the National Swedish Space Board, CCRS, and CSA.

Radarsat: The Canadian Radarsat program has acquired complete global coverage in the ScanSAR mode during its initial phase of operations (Mahmood et al., 1998) An associated PI research program was initiated to demonstrate applications, including applications to forest are research. Research to date has shown Radarsat to hold promise for mapping logging and deforestation in tropical and boreal forests (Shimabukuro, et al., 1997; Kux, et al., 1997; Ahern et al., 1997) and for monitoring macrophyte growth in tropical reservoirs and floodplains (Costa et al., 1997). The frequent, wide area coverage available, unaffected by cloud cover, makes Radarsat attractive for filling in gaps in optical data caused by persistent cloud cover.

Landsat - 7: Landsat - 7 has become part of NASA's Mission to Planet Earth, and will be launched in 1999. It will fly in formation with EOS AM-1, enabling researchers to obtain low and high resolution images of the same area on the same day. MODIS data will enable atmospheric corrections of Landsat data to become operational for the first time in the program.

The data price will be much lower than the previous commercial price, and there will be no copyright restrictions. A data acquisition strategy which incorporates data needs, local phenology, and local climate and weather, combined with an on-board solid state data recorder capable of recording 200 scenes per orbit are expected to provide global coverage which is more consistant and useful than in the past. These changes in policy and strategy will make it much easier to achieve the GOFC objectives.

SPOT Vegetation: The SPOT 4 mission will pioneer the simultaneous acquisition of wide area, low resolution data supplemented by pointable high resolution data, and stimulate the development of suitable data acquisition strategies (Phulpin and Gentet, 1993). The addition of a spectral band in the 1.6 to 1.8 μm region on both sensors will provide added sensitivity to damage from fire, insect attack, air pollution, and other causes (Vogelmann and Rock, 1988; Ahern et al, 1991).

FIRS: In 1994, the Joint Research Centre of the European Commission, in cooperation with the Commission's Directorate General 6 FII.2 and EUROSTAT, launched the FIRS (Forest Information from Remote Sensing) Project (Folving et al., 1995). The objectives of the FIRS Project are:

1. To contribute to the development of a unified European forest information system;
2. To develop methods for providing both sectorial (i.e. production-related) and environmental (i.e. ecologyrelated) forestinformation, in the form of both statistical and mapped data;
3. To use remotely sensed data as a major source of information;

4. To use geographical information system (GIS) techniques to reduce the cost of data collection, data handling and data distribution;
5. To assist in the development of new methods for forest data collection and communication systems using space technology.

The project comprises two main parts. The first part consists of three Foundation Actions: 1) Regionalization and stratification of European forest ecosystems; 2) Design of a system of nomenclature for European forest mapping and 3) Compilation of a European geo-referenced forest data directory.

The second part is divided into six major Application Themes: 1) European forest statistics; 2) European afforestation monitoring; 3) European forest mapping; 4) European forest monitoring; 5) European forest modelling and 6) Mapping and monitoring of European grassland and non-forest natural vegetation.

Remote sensing use for forestland management at the national and sub-national level Since the launch of Landsat 1 in 1972, numerous national and subnational agencies have experimented with and used earth-observation data for a variety of forestland mapping and monitoring activities.

In most cases, the user agencies have been satisfied with the results as long as the work was competently performed and the user expectations did not exceed the the information content of the data. Nonetheless, it is generally agreed that earth-observation data has not fulfilled its expectations. Impediments to wider use include:

"High cost of earth-observation data. The new U.S. Landsat data policy is expected to help reduce this problem;

"The technology seems very complex and specialized to the un-initiated. Many consulting companies have sprung up to aid with the selection and use of the data, but this creates increased costs for the user;

"Earth-observation data provides less-detailed "top-down" information, while forest inventories have

traditionally used a "bottom-up" approach with detailed information obtained from ground surveys and aerial photography. There has not been enough effort put into developing methodologies to incorporate the best features of both approaches;

"It is too difficult for unitiated users to know which E-O data sources are most suitable for their needs, and to locate and procure suitable data. The CEOS Working Group on Information Systems and Services (WGISS) is leading an international effort to address this problem.

Regional Activities: Regional Activities have been initiated in recent years which could be coupled to the program outlined here for in-country work. Within the IGBP System for Analysis, Research and Training (START) a number of regional activities focusing on the land cover change question have begun. Most notable is the Southeast Asian regional activity which is directly linked to the EPA Pathfinder project in that region, as well as the aforementioned IGBP LUCC and High Resolution Data Exchange projects. START/Southeast Asia has initiated a Land Use and Cover Change (LUCC) program in two components: (a) a complete region-wide analysis of deforestation trends in the region, and (b) specific case studies established in four participating countries: Thailand, Malaysia, Indonesia, and the Philippines. The case studies, conducted primarily by local experts and scientists, provide a prototype example upon which to build in-country work. The Thailand Country Study is already a component of the START/LUCC activities in Southeast Asia. In Europe, the Centre for Earth Observation (CEO) at JRC/Ispra has stimulated numerous projects to develop and implement practical applications of remote sensing technology for forestland management.

State of the Art

Technical Progress: Considerable technical progress has been made during the past five years in assembling large data sets from both high resolution and coarse resolution sensors, and in extracting information from these data sets. Increasing computer capability to move, store, and track massive amounts of data has been an essential enabling technology. Most of the coarse resolution studies have used AVHRR data. Researchers have developed methods for geometrically correcting the data and creating single-date mosaic images and multi-date composite images which are relatively free of geometric errors and radiometric artifacts. The TREES project and the IGBP 1 km Land Cover Mapping Project have both produced land cover classifications of large fractions of the world, while other efforts have produced land cover classifications at the national or continental level. Validation efforts are now underway for all of these products.

The Landsat Pathfinder program has enabled several organizations to develop methods to collect and assemble datasets of 30 m resolution data covering areas as large as the conterminous U.S. and the Amazon Basin. This experience has also pointed up the difficulties in collecting and correcting data from the existing Landsat data archive, and stimulated improvements for Landsat-7. The lack of accurate base maps to correct data of resolution below 100 m, coupled with the lack of high resolution DEM threatens to limit the accuracy with which high resolution data can be corrected and superimposed.

A significant number of researchers and operational users of both coarse and high spatial resolution data appear to be converging on a common approach to digital classification of multispectral/multitemporal optical data: numerical clustering of the data into a large number of data clusters (typically 50 - 100 clusters; sometimes called hyper-clustering), followed by merging and labeling the data clusters into desired thematic classes. The merging -labeling step is carried out by interpreters who are knowledgable about the land areas being studied, and seems to combine the best characteristics of digital and visual classification at the current time. This process is still very laborious, and digital image classification will be an important consideration for GOFC. Research with L-band (JERS-1) and C-Band (ERS and Radarsat) SARs has built upon experience gained with airborne SAR missions and the SIR-C/XSAR mission to provide confidence in the use of SAR for GOFC. While the thematic content of single-band orbital SAR is not as great as that of optical sensors, the ability to cover large areas in a preplanned, systematic manner was recognized as a very important operational advantage at the Ottawa workshop. Techniques for systematic geometric correction (except for terrain correction, where DEM data can be a problem) and radiometric calibration of SAR data are becoming highly automated. The Global Rainforest Mapping Project has gained considerable experience in assembling JERS-1 mosaics of large areas, and the routine radiometric calibration of the data is so good that no further radiometric processing of the mosaics is necessary.

Institutional Progress: The IGBP and the Global Observing Systems have created well-functioning international bodies to ensure ongoing dialog between the producers of Earth observation data and the scientific community, addressing the very difficult global issues facing humanity. Most of the projects described in section IV A have created temporary or permanent inter-institutional linkages which have functioned well.

The World Forest Watch, the FAO Forest Resource Assessment, and the GAP Analysis Project have demonstrated different ways to extend Earth observation technology to a multitude of users, particularly those involved in natural resource management planning. However, at a higher national and international level, the link between the information which can be obtained from Earth observation data and the formulation of policy has not been as well established.

Gaps and Lessons

Some common lessons have been learned from the above-described projects.

Data Acquisition, Access, Pre-processing and Management: A systematic global data acquisition plan is necessary and requires the contribution of multiple ground receiving stations. Data have to been processed in standard format and well-documented in the catalogs. A comprehensive metadatabase including all the high spatial resolution sensors'images would be an important improvement for data access. Within the project, an automated information and archive management system is required. A high resolution DEM is needed in hilly and mountainous areas to orthorectify spaceborne data. This is particularly true for SAR data which is more severely affected by the effects of slopes and elevation variations.

Product definition: Achievable products, not overly ambitious, must be defined. They must be explicitly targeted to specific user requirements. Legend definition and analysis design require the contribution of a large panel of experts, users, and potential users near the very beginning of the project. It is essential to document explicitly the forest classes and propose a correspondence between the sensor-based classes and the ecological classes.

Data analysis: Operational data analysis must be based on existing and proven methods rather than relying too much on R&D. Data analysis methods can combine visual interpretation and digital classifications. For land cover change, a control mechanism is needed in order to check the consistency between the multi-date analyses and avoid excessive error propagation. Change information may be more reliably obtained by classifying twodate datasets than by comparing two single-date classification.

A multi-resolution approach will allow the Project to exploit the complementarity between high and low resolution data: a sampling scheme of high resolution is optimized using low resolution data as a stratification, while high resolution data can correct the areas derived from low resolution data.

Particular attention will be given to a validation schemes. Products' validation may be performed through comparison with immediate finer resolution data, i.e. fine resolution products can validate low resolution products, field observations can validate fine resolution products. National experts should contribute as soon as possible to the product validation.

Product Distribution: Open access to final products and validation sets is a key point for the success of the project.

Framework

Strategy

The Pilot Project is anticipated to have an overall lifetime of approximately five years, with initial test products being available within the first three. It anticipates several categories of outputs: basic scenes, specific time frame products, and change over time products.

(1) Raw data (Level 0) and corrected scenes (Level 1B). These represent composites of the basic raw data for all sensors covering all forest land areas of the globe. Data products at this level would enable users to undertake projects to contribute to preparation of global products identified in Table 3, as well as products that fulfill specific national and/or regional needs above and beyond those identified in Table 3. This level of product preparation is viewed as an absolute minimum in terms of project success. Optical sensors will provide high thematic resolution and temporal resolution at low spatial resolution, for wall-to-wall coverage, detection of drastic events, and forest seasonality monitoring.

(2) Specific Time Frame Products. This is the preparation of products of global forest cover and characteristics identified in Table 3 for specific years or time periods. These data sets essentially would represent the status of forest land cover and forest characteristics at specific points or windows in time (ie T1990, T1995 or T1980-90 or T1990-95).

(3) Change Products. This is the preparation of global forest cover and characteristic products identified in Table 4 that show the change in these parameters over time. (ie. T1995-T1990). Most of the data acquisition issues follow directly from the primary project objective. Test products from the design and prototype phases (see Section VIII) are anticipated to be largely regional, and based on existing, available data sets. However, near the end of the prototype phase, the project will begin the production of derived products from new acquisitions. Thus, global

and regional data sets are required from the present generation of spaceborne imaging optical and microwave sensors and their near-term successors, in conjuction with concurrent in situ data. The CEOS members will need, to coordinate their respective strategies for the related future global-scale satellite acquisitions in order to acquire such data systematically and effectively. Particular organisational efforts may be required where an international network of ground receiving stations is involved.

For efficiency, the acquisition strategy should exploit the complementary capabilities and information content of optical and microwave sensors to address specific thematic issues on appropriate time scales: Optical Sensors providing both high spatial and thematic resolution, based on well-established methodologies, under suitable acquisition conditions.

Microwave Sensors providing all-weather acquisition capability, at high spatial resolution and frequent revisit rates, but low thematic resolution, and utilising analysis methodologies which will necessarily evolve throughout the project.

Validation plans will be formulated and implemented as feasible so that users of both raw data and derived products will know their accuracy and limitations. Feedback will be solicited from users to refine derived products in order to maximize their utility.

Simple Global and Detailed National Products

It is anticipated that the global information needs in most cases are less detailed or specific than national or local needs. Hence, flexibility in data resolution and product specifications is both a requirement and a benefit of this project. Preparation of products at three levels provides this flexibility. Where specific user needs are more specific or detailed, the Level 1B (see Appendix 1) data should allow users to develop products at the level specific to their needs and requirements. At the same time using a common agreed upon global format or classification should allow more detailed national and regional information to be rolled up to meet the basic global requirement. Hence, the more simple global needs can be met while at the same time providing the flexibility to meet more detailed national and local needs and requirements. It is expected that this project will stimulate much activity at the national and sub-national level for practical forestland management, above and beyond that related to the global portion alone.

Product Identification

CEOS Level 0 and 1B Products: The Appendix presents the lowest level products, characterized as level 0 and 1B CEOS products. These products are basically "raw" (level 0) data or data to which calibration and geometric processing are applied (level 1B). They will not be discussed further.

CEOS Level 2 Products: Table 3 contains level 2 products that are obtained from level 1B data, but are based on more complex processing techniques, like classification algorithms or functional dependencies, and may require in situ or ancillary information to be developed. (Global Climate Observing System, 1997.)

Fine Scale Land Cover: Products derived from high spatial resolution data are georeferenced data sets describing the predominant forest types that can be used as a baseline inventory for long term studies of land cover and land cover change. The temporal coverage of this data will be less than that of mid-low spatial resolution land cover products.

The total global number of classes could reach 50 or more. They will include various wooded ecosystems : temperate and tropical forests, savanna, and others. These products may be derived from various sensors.

Coarse Scale Land Cover: Products derived from low spatial resolution data can be updated more frequenty than the fine scale products due to the sensors'data acquisition cycles and lower data volumes. The total global number of classes could reach 40 or more. However, observational constraint imposed by the coarser resolution will prevent the detection of changes affecting areas of less than several km2. Areas affected by large wildfires will be type of change most suitable for frequent update using coarse resolution sensors.

Fires: Fire is a primary force of vegetation change/renewal in some ecosystems. It is important to monitor fires in the context of a forest monitoring program with respect to their role in global trace gas emissions, land conversion, ecological disturbance regimes, and land management. Data products will consist of fire occurrence and burned area, providing the location, timing, and areal extent of burning. A thermal channel product will provide hot spots which may indicate forest fires.

Harvest, regeneration, and afforestation: It is important to distinguish between land cover change due to clear cutting and selective logging. Harvesting, regeneration, and afforestation are some of the most significant anthropogenic changes that occur in the forest. These often occurs in similar locations from year to year.

Biomass: This product will largely depend on in-situ field measurements and other ancillary data. Earth observation data cannot provide direct measurements of biomass, but may be used to extend local estimates to larger areas, with varying degrees of confidence.

LAI, FPAR and PAR: These products provide the necessary information to derive net primary productivity (NPP) using process models at the landscape level (Running et al, 1989; Liu et al.., 1997). NPP is an essential component of all carbon budgeting calculations.

FPAR, the fraction of photosynthetically active radiation absorbed by a vegetation canopy is readily measured from space on clear days from observations of the upwelling radiance above the canopy, since the downwelling radiance (PAR) can be readily estimated. Daily observations with polar-orbiting meteorological satellites are well suited to these observations. Progress has recently been reported in extending these techniques to cloudy days using frequent observations of upwelling radiance (Li and Moreau, 1996; Moreau and Li, 1996).

Geostationary meteorological satellites provide the most suitable observations of PAR, which can exhibit rapid temporal variations under conditions of variable cloud cover.

Leaf Area Index (LAI) is required to calculate the efficiency with which APAR is converted into stored carbon. LAI can be estimated from observations of FPAR, but knowledge of land cover is also needed to produce estimates of adqueate accuracy.

CEOS Level 3 Products: We propose to limit the land cover change measurements to regions at high risk of disturbance, because land-cover change only affects part of the global forest. This stratification can be achieved by starting with coarse scale indicators (fires, population, logging concessions, etc.). The land cover change measurements can be based on sampling strategies or exhaustive mapping in sensitive regions.

Data Bundles For Users

The Pilot Project will also undertake to enhance the value of derived products for particular user communities by providing them as part of a data bundle. These bundles are collections of data and derived products which particular programs or users can use most effectively to fill a particular need, thus cutting down on the provision of data that are non-essential to that need.

The example below shows how the data bundling might work to meet the requirements of the Framework convention user community. A bundle is proposed for emission inventories, which focuses on data for landcover change and biomass burning. Multiple sensors are required to produce these complete packages, and CEOS can therefore play an important role by providing coordination and cooperation in acquiring and processing data from different sensors, platforms, and programs.

Here we provide an example as it might be done for emission inventories, and carbon cycle research. An Emission Inventory data bundle is shown below.

Input Data Sources from Archive

Input Data Source/Sensor Sp. Resolution Temp. Res. etc

Deforestation, decadal deforestation, annual

SPOT, TM, IRS 10-50 m 5 yr....

SPOT, TM, JERS 10-50 m 1 yr

Logging JERS, Radarsat 10 - 30 m 5 yr

Acquisition Strategy

An acquisition strategy must be developed to ensure the necessary data are acquired to produce each of the required products. This is particularly true for the data bundle products, where data from multiple sources must be combined to produce the required product. It will be necessary to create an organization to continuously monitor data acquistion and modify acquisition plans to ensure that complete coverage is obtained.

Data Processing Bundle

This data bundle provides a package which includes the 5 data products necessary to make an emission inventory as well as the bundle of requisite satellite sensors; it provides a framework for plugging in datasets as objects, and provides a framework to define CEOS partners participation. It also couples the acquisition strategy directly to the products and the user/ science requirements. Finally it couples a data processing bundle which include the algorithms, data distribution mechanisms, formats, and information systems necessary to produce the data archive and the products.

METHODOLOGIES

Sensors

A preliminary list of sensors relevant for the project has been identified. The focus of the Pilot Project is on those systems that are either currently in orbit, or are funded and in the development cycle for launch within the next several years. There are clearly many other missions that might become available within the next decade, but they are beyond the scope of the current Pilot Project.

(1) Optical coarse resolution sensors (1km - 200m)

These sensors provide high thematic information with high temporal sensitivity, but at coarse spatial resolution for seasonal evolution and monitoring studies.

Exisiting: AVHRR,ATSR, Resource MSU-SK, WiFS

Near-term : AATSR, SPOT Vegetation, MODIS, GLI, MERIS, MISR,POLDER,SEAWIFS, CEBRS WFI Camera

(2) Optical Fine resolution sensors (200m - 15m)

These sensors provide land cover and land cover change information with great spatial and thematic detail. However, the data volumes are very large, and it is not yet possible to automate the classification process completely. This necessitates a long interval between land cover updates. Best use of resources can be achieved by revisiting areas of rapid change more frequently.

Existing : SPOT/HRV, IRS/LISS II, Landsat 5/TM, MOS/MESSR.

Near-term: SPOT4/HRVIR,Landsat 7 ETM+, ASTER, ADEOS AVNIR, CEBRS CCD Camera, Resource MSU-SK.

(3) SARs (fine-resolution 150 - 20 m)

SAR (Synthetic Aperture Radar) sensors can acquire data through clouds and are needed in areas of nearly continuous cloud cover, such as tropical rainforests and some coastal areas. They can provide information on land cover change.

Existing: ERS1,ERS-2,JERS-1,RADARSAT

Near term: ASAR,ALOS,SRTM, Lightsar?

(4) Hyper-resolution Optical sensors (0-1m)

This data is expected to be available commercially during the course of the GOFC project. A very limited set of data from these sensors could contribute to validation.

Data Access

The procedural aspects of data access deal with metadata formats and data browsing and query systems on the one hand,

and the delivery elements of timeliness, media and means on the other hand.

The Pilot Project will investigate a unified system of metadata maintenance and integrated mechanism of cataloguing, querying and ordering. Integrated systems may be ensured by remote sensing data distribution and access networks such as EOSDIS and CEONet currently under development. Although a unified system of metadata maintenance may be sufficient to provide the first level of information on a data set, the availability of a quick-look facility at the browsing and query stage may be highly desirable to know the quality of the data before these can be ordered.

The delivery elements that need to be taken into consideration are the turn around times covering the data ordering, data reception/archive retrieval, data processing and data transmission cycle. Agreements on standardization of delivery medium are important in view of the variety of strategies adopted at various output centres. The means of delivery are wide ranging, but need to be verified for each of the missions participating in the project.

Data Pre-Processing and Management.

All data provided for the Pilot Project should be pre-processed in a standardized way in order to maximize the amount of automation that is feasible.

Concerning the data management, the lessons learned with the CEOS initiative 'High Resolution Data for IGBP' show that it is crucial that there be clear responsibilities and accountability for data management and distribution. Tools will be required to compile the metadata to make up a Global Forest metadata base. A tool for inventory tracking has to be designed and implemented at the management level and probably also in each producing center.

The project archive could be distributed among the various producers and consulted through a interoperable catalogue. A documentation system (web site, CD-ROM, etc.) must be created.

Analysis

The aim of the analysis is to produce global datasets of forest cover and attributes.This will involve integration of data from a number of satellite instruments to provide global coverage at appropriate resolution and timescales. A key issue to be addressed by this project is therefore harmonisation of methods and results to ensure global consistency and comparability.

The outputs desired are anticipated to require two forms of analysis. The first is direct production of a thematic classification of forested areas, while the second requires modelling of the data to derive information on forest functioning.

Thematic Classification: Methodologies for thematic classification of forest from coarse resolution satellite information (eg TREES) and fine resolution (eg Landsat pathfinder) have been an active research area for many years. Results from microwave seonsors have also demonstrated distinguishing forest, inundated forest, and non/forest with high temporal coverage.

The number of classes envisaged and the useful spatial resolution implies that considerable analysis of fine resolution satellite data will be employed. It is anticipated that data from a number of fine resolution sensors will be required in order to obtain adequate cloud-free coverage. A key issue to be resolved is how to combine the results of classifications resulting from different instruments, particularly with respect to analysis of change detection.

The coarse resolution instruments have the capacity to provide information at higher temporal scale, giving more opportunity to use phenology in class seperation, and to optimise the acquisition strategy for data processing. For the production of thematic classes, ancilary data will be required in order to identify categories such as dominant tree species. An important contribution will be to direct and stimulate research leading to greater automation of image classification.

Derivation of information on function: The estimation of the net primary productivity (NPP) requires a model of ecosystem function. Driving variables in addition to the classification information are LAI, FPAR as well as ancillary data on downwelling PAR, temperature etc. Information for forest functioning models is required at a much higher temporal resolution than the classification (at timescales of one hour to ten days).

Accordingly, while an accurate classification of land cover is needed, these applications will mostly require coarse or moderate resolution data. Currently, a number of models exist specified for different ecosystems, and there is not a consensus on best methods to invert the necessary parameters from remote sensing observations.

It is recommended that accuracies from different platforms are made comparable, and that assimilation techniques are developed similar to those routinely employed in meteorology. Further ancillary data are required to both drive and validate model function, and to derive parameters such as biomass.

Ancillary Data

Several types of ancillary data will be needed for the pilot project both for data pre-processing, satellite data analysis and product validation.

For example data processing of the coarse resolution data will involve the use of coarse digital terrain data to provide first order atmospheric correction and improved geolocation for coarse resolution data sets used in classification.

The existing global 1km topographic data set will be adequate for meeting these needs. Similary a common land water classification and mask will be needed for the coarse resolution data products. An enhanced land water mask for use with coarse resolution systems, derived from a combination of microwave and moderate resolution sensing systems available in the near-term would provide a desirable improvement over the current data base derived from the best available map sources. The development of a standard 'community'land water mask derived from multiple sensing systems would be a useful CEOS contribution to this activity.

For the individual sensor products and the multi-resolution integration there will be the need for the data products to be presented in a common geographic coordinate scheme. This will have to be developed early in the project. For high resolution, the availability of a new high spatial resolution satellite derived digital terrain data towards the end of the project will provide the possibility of improving high resolution analyses in areas of complex terrain.

The use of information from non-satellite sources will play an important role in the development and evaluation of the satellite derived products. This information will be in the form of existing maps, air photos, and field data. In some cases these data will be used as part of the classification process to train the classifiers, in other cases they will be used to test the accuracy of the thematic products.

For example as part of the high resolution accuracy assessment, sample geolocated field data will be needed. The development of the sampling strategy will need to be developed as part of the design phase of the project. In situ data on biomass is part of the product suite. In this context it is seen as a primary data source rather than an ancillary set. The combination of satellite data and in-situ biomass data will provide the necessary spatialisation needed for the biomass product.

Product Validation

Product validation will be critical part of the project, providing the user community with a level of confidence concerning the results provided. It is important that the accuracy of the various products be stated quantitatively and explicity. A set of procedures for validating coarse resolution global land cover data sets is being piloted by the IGBP-DIS global land cover project.

The procedures include a rigorous statistical core sample supplemented by a series of confidence sites where existing information are readily available. Regional scientists provide an important part of the product validation. High resolution products are used to validate the coarse resolution products.

Validation of high resolution data products will be performed by a combination of hyper-resolution data (3-5m) and precisely located field data collection. We expect that CEOS affiliates will play an important role in product validation.

The relationship of the global products and associated validation activities of this project to the GTOS/TOPC GHOST network of field sites and the relevant product validation activities of the different space agencies need to be established early in the project.

For example forest test sites within the GHOST network could play an important role in the pilot project validation. The relationship of this pilot project validation activity with respect to the on going CEOS calibration/validation activities also need to be determined. It is important that the validation data be packaged in a usable form and made available along with the products and their associated accuracy statements. A mechanism should be established to allow for users to provide feedback on product accuracy.

Product Distribution

Products should be distributed to user communities including scientists, policy makers and other operational users as promptly as possible in different ways and through different media. In particular, reporting to policy makers is of great importance. Several different scales and levels of data should be provided such as global, continental, national, and if possible sub-national so that policy makers can evaluate the state of the environment in their particular areas of interest, thus making informed decisions possible. In some cases products will be distributed with appropriate interpretation to policy makers so that they can understand the products in the context of the information that they require.

The Internet is a very efficient way to distribute derived products and information about products and data to users.

However, other media will also be taken into account in order to satisfy users without reliable access to electronic networks. Alternative media includes CD-ROMs or other digital media, and paper prints, maps, and reports. Some users will require access to the actual underlying satellite data. In most cases, network delivery of such data is difficult, and tape or other media may be preferable.

IMPLEMENTATION ISSUES

Management

Oversight by CEOS: CEOS will provide programmatic and technical oversight to the Pilot Project. This function includes evaluation of Pilot Project performance, and an evaluation of the degree to which the Pilot has been successful in addressing the needs of targeted user communities.

Project Team: The Project Team will come initially from the CEOS members and affiliates, using whatever national or programmatic mechanisms each institution has at its disposal. In addition, early in the life of the Pilot Project, a concerted effort will be made to identify and incorporate team members from national and international forestry and environmental organizations, so as to ensure that there is broad representation from both providers and users of data.

The Canadian Forest Service has expressed an interest in developing a national forest inventory based on earth observation data, thereby serving as a challenging test case. The first tasks of the Project Team will be to form a Design Team to specify the technical details of implementation for the first phase of the project.

Data Issues

Satellite Data Access and Redistribution: The participants at the Ottawa workshop agreed that GOFC will benefit from liberal data access policies, that is, data which are available at minimal cost and with minimal restrictions on re-distribution. However, they recognized that satellite data policies vary between CEOS members, and that members may insist that data made available for GOFC be subject to their established data policies. All scientific members of the project require access to the data for analysis and publication of results. The project team will also share satellite data for which additional processing has been performed.

Ancillary and in Situ Data: All data will be calibrated. Ground calibration data will be made available within the project team. All products generated also require the inclusion of ancillary data for analysis, such as topographic data. For some regions there will be available additional geographic information layers. Each member of the project team will have full access to these additional ancillary data sets. The project team will maintain a catalog of all data acquired, and its distribution.

Analytical Techniques: The products to be created from IGOS project activity should have clear statements on their accuracies. The products should be unbiased estimations of forest properties.

The project team will fully document the analytical techniques used for each product. These analytical techniques will likely vary from region to region, and consistency of results on a global basis is an important goal of the project.

Communication: The project team members will communicate electronically on a free basis. A variety of methods for electronic communication with CEOS, the user communities, and the public will be developed. In addition, there are expected to be many major publications in the open, reviewed literature, and internal publications and reports for the CEOS members and affiliates themselves.

Public Communications: Given the high public interest in the state of the forests and environment, it is essential that results of forest observation be widely communicated. Press releases should be prepared for publications which are widely distributed, and to television and radio networks. The high visual content of our information is appealing to the public. The project team should also create CD-ROMs with integrated data and products which can be provided to the public and to educational institutions. The project team should give consideration to the creation of videos about regional and global forest cover and changes in forest cover.

Additional Research Opportunities

The Pilot Project is not perceived as a primarily research-oriented exercise. It is an experiment in operational capabilities. At the same time, it is clear that there are many research opportunities which may benefit project planning and execution. Some of these are outlined below.

Hyperspectral: Optical sensors with large numbers of channels will shortly be launched. Sample data sets from space and airborne platforms will be available.

Very High Spatial Resolution: Several commercial spacecraft are planned which will have resolutions of the order of one meter. There are methods for recognizing individual trees from such data. Such data can also be used to assess selective harvesting, partial cuts, and regeneration, and provide information on site characteristics.

Polarimetric and Interferometric SAR: An important advance in microwave sensors is the creation of SARs with multiple frequencies and full polarization. Polarimetric SAR data sets and interferometric SAR data sets are available for sample areas.

Lidars: In addition to their contribution to atmospheric measurement, lidars can be used for measuring tree heights and for creating DEMs. During the lifetime of the Pilot Project, sample lidar data sets are expected to be available.

Recommendations from the FAO representative with respect to the workshop process and resolutions.

Although the need for a global integrated (source) data set specifically designed for the observation of forest resources is fully endorsable, in view of the potential benefits to all user groups and on-going forest monitoring programs, the specifications of derived global products should be the result of a wider consultation process that includes representatives of all user communities identified under section III B of the Workshop Report. This is particularly important for the product based on high resolution data which would provide crucial information on land cover/forest cover state and change at national and sub-national level with consequent important strategic contribution to national planning operations.

The initial phase of the proposed pilot project will include a mechanism of thorough user consultation in order to define their priorities, propose suitable mechanisms of participation/ consultation and better define product specifications.

Such consultation should include, as a minimum, forest management/planning institutions' representatives from major "forest" countries (Russian Federation, U.S.A., Brazil, Indonesia, Democratic Republic of Congo [former Zaire], etc.), and International Policy and Operational Institutions (FAO, UNEP, UNDP, World Bank, etc.).

SCHEDULE OF ACTIVITIES

Overview

The Ottawa workshop in July 1997 has provided the first kick-off for the GOFC project. Participants in the meeting represented a wide spectrum of international experiences with forest-related programs. There were also representatives of several potential user communities. However, all participants recognized that this was only a beginning. There are many strategic and technical issues that remain to be worked out before the GOFC can be regarded as fully implemented.

Project Phases

There are three main phases that have been identified, with several follow-up activities.

(1) Design Phase (1 year). The first year of the GOFC, beginning with project approval, will be devoted to a series of detailed design studies. These would include workshops on specific technical issues, e.g. to analyze

the acquisition strategies for existing and planned remote sensing missions. Design studies might also include paper studies by selected study teams, or comprehensive literature reviews, and specific methodological intercomparison studies, e.g. studies comparing different classification methodologies using identical data sets. A specific need during the Design Phase will be a broader outreach effort to potential user communities, and specific planning for an eventual transition from the pilot phase to routine operational status for the GOFC. The ultimate goal of the Design Phase is to produce a detailed implementation plan that incorporates all the design issues. Deliverables will include a series of publications, project reports, and the data sets from intercomparison studies.

(2) Prototype Phase (2 years). This phase can be regarded as the initial implementation phase. This is the period during which the details of preprocessing, data handling, analytical procedures, etc., which have been planned during the Design Phase, will actually be tested for their operational practicality. Current experience shows clearly that design studies, while necessary, cannot anticipate all the eventualities of having to cope with large data sets in a routine manner. The Prototype Phase anticipates this difficulty, and provides the opportunity to work with large data sets in a mode analogous to what will be necessary with new global acquisitions. During the Prototype Phase, a series of regional pilot studies will be performed with existing data sets, which will be expected to produce the first sets of products and product bundles identified in previously, but for periods from the late 1980's and early 1990's. During the latter part of the Prototype Phase, new data from remote sensing systems should begin to become available (e.g. the first few months of MODIS data). These data could then be used to ensure that all the data handling systems for GOFC are operational.

(3) Execution Phase (2 years). The Execution Phase builds on the experience gained through the Prototype Phase. During this last phase of the GOFC, new data acquisitions are incorporated into the GOFC analytical system, analyses performed by the project team, and new products generated and distributed. The acquisitions for the Execution Phase will likely begin during the Prototype Phase, because of the time necessary for processing and analysis. Products and product bundles will be generated for the late 1990's and early 2000's.

Validation studies will also be begun during the Execution Phase, although it is not clear that all of them can be completed during this time. However, the conduct of validation studies will be part of the overall evaluation follow-up activities.

2

Fucus Vesiculosus in the Atidal Baltic Sea

INTRODUCTION

Fucus vesiculosus L. is the only widespread, large perennial seaweed in the Baltic Sea where it dorninates the biomass (Jansson & Kautsky 1977, Kautsky 1988) and is the base of the most diverse hard-bottom community with some 30 species of macrofauna and epiflora (Haage 1975, Kautsky et al. 1992). Reduction in the distribution of the Baltic E vesiculosus population (Kangas et al. 1982, Lindvall 1984, Makinen et al. 1984, Ronnberg 1984, Kautsky et al. 1986, 1992) has made it urgent to study factors governing the reproduction, recolonization potential and survival of this seaweed. Studies on the age and size structure of E vesiculosus populations from a wave exposure gradient (Kalvas & Kautsky 1993) and those in a pulp mill effluent area (Kautsky 1992, Kautsky et al. 1992) showed that factors like wave exposure, pollution and grazing pressure are all important for the size structure and possible reproductive success of new gerrnlings. The mean size of F, vesiculosus plants at first reproduction has been shown to be about 25 cm in the area around the Asko laboratory (Stockholm University), Sweden (Kautsky et al. 1992).

In the northern Baltic proper, reproduction of Fucus vesiculosus occurs mainly in June and July, while Carlson (1990) reported a longer reproductive period in the southern part of the Baltic Sea, with 2 peaks. The initiation of receptacles in this dioecious species starts in mid-October in the northern

Baltic proper after the plants have been exposed to a short-day stimulus for 2 wk (Back et al. 1991). The receptacles remain dormant over winter until the ice-melt in April (Back et al. 1991). In ripe condition they contain oogonia with 8 eggs each or antheridia with 64 spermatozoids.

Rhythmicity of varying period length has been reported for reproductive activity, i.e. gamete release, in several algal populations, e.g. in different Derbesia spp. (Halicystis) gametophytes (see Ziegler Page & Kingsbury 1968), in Ulva lobata (Kiitzing) Setchell & O Inter-Research 1994 Resale of full article not permitted 196 Mar. Ecol. Prog Ser.

Garder (Smith 1947), Enteromorpha intestinalis (L.) Link (Christie & Evans 1962, Pringle 1986), Laminaria saccharina (L.) Lamour. (Luning 1981), in the dioecios brackish Fucus cernoides L. and in different red algae (Umamaheswara & Kaliaperumal 1987).

Although observations on some of these intertidal seaweeds indicate the action of a lunar-related biological clock, there is little experimental evidence. In a recent review on the ecological perspectives of sexual reproduction in algae, Brawley & Johnson (1992) summanze the various factors which have been suggested to affect gamete release in macroalgae in relation to a lunar or semilunar cycle. In Dictyota (reviewed by Sweeney 1987) an endogenous rhythm has been demonstrated in which moonlight is suggested as one of the presumed timekeepers.

In some species, gametes are known to be shed in response to changes in light. In most such cases re- !ease occurs at daybreak (Zieglei Page & Kingsbury 1968, Burr & West 1970). In contrast, eggs in Laminaria spp. (Luning 1981) and gametes in Pelvetia fastigiata (J. G. Agardh) (Jaffe 1954) have been shown to be released at the onset of a dark cyde in laboratory experiments.

Most effects of illumination on gametes have been studied in the laboratory. In the field, gamete release has been observed in Ulva lobata in morning high tides (see Smith 1947) and in Fucus ceranoides largely in daytime high tides (Brawley 1992).

Other factors that might be required or that stimulate gamete release are a change in salinity (Brawley 1992), replacing of the culture medium with fresh medium (La Claire & West 1978) or specific temperatures (Bacon & Vadas 1991). However, factors shown in laboratory studies to be of importance for gamete release do not always appear particularly important in the field.

For example, Pelvetia fastigjata may shed gametes when receptacles are placed in darkness after photoinduction (Jaffe 1954), or gamete release by Sargassum muticum (Yendo) Fensholt may be influenced by temperature in laboratory studies, but the semilunar cycle or other factors appear to have more influence on this event in the field (Brawley & Johnsson 1992 and references therein).

In earlier laboratory experiments with Baltic Fucus vesiculosus (Andersson et al. 1992), we observed a tendency for gametes to be released in the evenings at fortnightly intervals.

It is most likely that Fucus vesiculosus immigrated into the Baltic Sea from the Atlantic less than 3000 yr ago, when salinities became tolerable. At the same time it moved from a tidal to an atidal environment and from a littoral to a sublittoral habitat.

There are no reports about tidal or fortnightly rhythmicity in seaweeds living in atidal environments such as the Baltic Sea, and it would be of great interest from a genetic as well as from an ecological point of view if such a rhythm could be documented.

MATERIAL AND METHODS

The experiments in the Baltic Sea were carried out at the Asko laboratory (Stockholm University) in the northern Baltic proper. Although there are extremely small tides of only 0.05 m in the Baltic Sea, the water level may fluctuate irregularly by some meters over the year due to changes in barometric pressure and onand offshore winds. The salinity was low and stable in both the field and the laboratory experiments, about 6.5%0.

The experiments were performed in 1992. Male and female Fucus vesiculosus individuals were marked in the field and receptacles were collected from these plants each day. Twelve ripe, even-sized female and male receptacles were collected, with a mean (+ SD) number of conceptacles per receptacle of 169 + 12. One male and one female receptacle were piaced together in small (20 ml) transparent plastic containers with 10 m1 of seawater.

Three receptacles were replaced each day. The plastic containers were put in the field, floating in a transparent plastic bag at the surface within a larger F, vesiculosus stand. Temperature was measured in the F. vesiculosus stand as well as in the plastic containers at the sampling time, when water was changed.

The water level fluctuation was recorded each day at 10:OO h. The experiment was run daily from 28 May (the start of the main reproductive season, 1992) until 16 July (near the end of the reproductive season).

The water was replaced daily at 10:OO and 22:OO h. Starting from 4 d before each full and new moon, water containing released eggs and oogonia was sampled at shorter intervals, usually every 4 h. Insolation, measured as kJ m-2, was continuously recorded and calculated for each 4 h and 24 h period.

The water samples were brought to the laboratory, fixed with Lug01 solution, and the number of eggs and oogonia released were counted in an inverted microscope at 200x. The total number of eggs released per receptacle was determined by multiplying the number of oogonia by 8 after which the mean number of eggs released per conceptacle X 100 per day was calculated. The calculation per conceptacle was performed to reduce the effect of differences in size and number of conceptacles per receptacle.

The second experiment was performed under controlled conditions in the laboratory. The receptacles were not replaced during this experiment. They were kept in the same kind of plastic containers as in the field experiment, each container with 1 female in 10 m1 Andersson et al.. Gamete release in Fucus vesiculosus 197 of filtered seawater. In this experiment, 10 receptacles from 5 different female plants were collected, 2 receptacles from each plant.

The receptacles were kept cold on ice during the transportation (about 1 h) to the laboratory. Five receptacles, 1 from each plant, were placed directly into the containers. Five other receptacles were put to dehydrate on moist paper overnight (for details about treatment, see Andersson et al. 1992).

The water was sampled each afternoon and the number of released eggs was counted as described above. To eliminate the 'background noise' in egg release between peaks, the individual mean number of eggs released per day during the experiment (mEi) was calculated for each receptacle.

The individual daily reproductive activity (dAc) for each receptacle was calculated as the ratio of eggs released each day (dEi) to mEi. Mean reproductive activity was calculated as the sum of dAc divided by the number of receptacles. Values exceeding 1.0 were considered as shedding activity. Calculations were made for control and dehydrated receptacles respectively. Egg release was counted from 6 July to 29 July at a constant temperature of 11 * 1 "C and a constant light:dark regime of which mimics the natural light conditions during the reproductive period under which the field experiments were carried out. The receptacles were accidentally exposed to light for a few minutes during the night of 8 July. The material was treated statistically (t-test and linear regression) using the Statgraphics computer program.

GRACILARIA PRODUCTION AND UTILIZATION IN THE BAY OF BENGAL REGION

The Bay of Bengal Programme began its involvement with Gracilaria farming in 1983 when technical and financial support was provided to the Fisheries Research Institute of the Department of Fisheries, Malaysia. Technical trials based on spore setting and outplanting off Penang Island were promising.

Over the years there has been considerable interest in the agarophytic resources of India, particularly along the Ramanathapuram coast of Tamil Nadu. Research by various institutes'indicated culture potential and in consultation with the Department of Fisheries of the Government of Tamil Nadu, BOBP initiated a pilot project involving two villages in Ramnad district.

Concurrently, a similar project began in Sri Lanka together with the National Aquatic Resource Agency and Sarvodaya Shramadana Sangamaya ; a prominent NGO with wide experience in the country. Various technical problems arose during the implementation of the two pilot projects, but information on current experience with Gracilaria production and marketing in the BOBP region was not readily available. It appeared an opportune moment to gather together as much expertise as could be assembled in one place and assess the "state of the art" in agarophyte culture, processing and marketing, as well as the extent to which natural resources were being managed.

Consequently, BOBP in collaboration with the Government of Thailand, convened a regional seminar in Songkhla, Thailand, October 23-27, 1989.

All the BOBP member states sent representatives and in addition, scientists, technicians and business people came from Vietnam, China, Hong Kong, the Philippines, Sweden, the United Kingdom, and the USA. International organizations interested in seaweed production were represented by ICLARM, INFOFISH, NACA and of course, BOBP. Development strategists, scientists, commercial processors and technicians were among the 82 participants. The seminar was the first regional convocation of the seaweed industry. No international seminar had ever dealt earlier with a particular genus (although there has been some mention of the closely related Polycavernosa).

The topical organization of the seminar reflected an interest in applied aspects of Gracilaria, and included sessions on the status of culture, small-scale processing, marketing, and on the management of natural resources of agarophytes. Twenty-four papers covered these areas of interest, of which seventeen dealt with culture and processing.

Almost all cultured Gracilaria derives from ponds: open water systems are still quite experimental in tropical waters. However, open water culture is operating commercially in the temperate waters of China; such systems have also been developed in Chile's cool seas.

Tide level and pond design are particularly important to ensure good flow of seawater. Access to both fresh and seawater enable the farmer to maintain salinity at an optimal level for fast growth. Sites should have protection from strong winds, and muddy sand or sandy soil is preferable. Several issues arose during discussions: Paramount is the high cost of developing new ponds; the soil requirements for pond construction can restrict the spread of the technology.

There may be conflicts with mangrove forest conservation if new areas are developed, since such tidal forests are widespread in tropical Asia. Soil acidity could be a very serious problem in ponds established in mangrove forest soils, but reclamation of acid sulphate soils is a lengthy and expensive process. It is possible to involve small-scale brackishwater pond farmers in Gracilaria culture as demonstrated by private industry in South Sulawesi.

Gruciluriu may be polycultured with other seaweeds or with fish and prawns. Polyculture of Grucilaria and Eucheuma has been experimented with in Bali, Indonesia. Trials have employed small-scale open water technology. However, in pond culture, protocols for fertilization regimes for specific soil water types and for maximizing seaweed growth and agar quality are still lacking.

The best stocking densities for seaweed, fish and shrimp remain to be worked out, as does the amount of residual stock that should be left after harvesting in order to maximize production. All of the development has been done by trial and error and may be an area for applied research.

OPEN-WATER CULTURE : THE PROBLEM OF GRAZING

Rabbit fish (Siganus spp) are the main grazers and they can literally wipe out large seaweed farms. It appears that juvenile rabbit fish are largely to blame; seaweed farms located in shallow seagrass beds are particularly vulnerable, since this is their natural habitat.

It may be that open-water culture is only possible where natural stocks of the desired species of Gracilaria are found. In practical terms, this may not always be possible. Turbid lagoons may be preferable to more exposed sites. Results of culture trials in Songkhla Lake itself and from earlier BOBP work in Penang seem to bear this out. Fouling by unwanted epiphytes and invertebrates may seriously affect product quality. Open-water culture of Gracilaria has been commercially successful in the Caribbean (St Lucia) but such success stories are rare.

In Malaysia, land based intensive systems have been tested. While of technical interest, there would be little likelihood of such systems being taken up by fisherfolk because of the high capital cost and the level of managerial expertise required.

SELECTION AND SEEDING METHODS

Culture systems of whatever nature require seed stock, and a number of questions arose concerning selection and seeding methods. All commercial production in the tropics derive from vegetative propagation rather than spore-setting. Parent plants are used to seed substrates such as stone blocks planted in open temperate waters in China. Similar techniques were tested in Hawaii with limited results but further work with this approach might be warranted in the tropics. Hatchery spore setting in Sri Lanka has been successful on a small scale.

As agar yield and gel strength vary greatly with species and even possibly within species according to variety, strain selection becomes difficult. Until the confusion over the taxonomy of Gracilaria is resolved, selection of strains will remain difficult. If there is a high degree of adaptation to micro-environments, how transferable will the results of strain selection be to new environments? Are the best species being cultured? Given the number of known species of Gracilaria, it was noted that only three (G. verrucosa, G. tenuisripitata and G. edulis) are cultured in Asia.

SEAWEEDS FOR HUMAN CONSUMPTION

Seaweeds for direct human consumption are worth more than those destined for industrial use. The nutritional benefits of seaweed consumption were considered as a source of micronutrients (vitamins and trace elements). On the industrial side, problems are presented by variations in agar quality and composition as influenced by particular clones, light and nutrient conditions and the interactions of these factors.

The history of agar dates back to seventeenth century Japan. Consumption of seaweed as a fresh vegetable is probably an ancient practice; it is found today in places like Hawaii, Fiji, the Philippines, Indonesia and Malaysia. Home production of agar for puddings, soups and jellies can be found in Sri Lanka. Perhaps production for local consumption has scope for encouragement in some of the BOBP member countries.

WORLD AGAR PRODUCTION

World agar production is estimated at 7,000 to 10,000 tonnes annually, and about half is from Gruciluriu. Japan is the top producer and consumer, but depends heavily upon imported Gruciluriu to meet national demand. In the BOBP region, Thailand, Malaysia and Indonesia are major importers of agar. In India, estimated annual production is 75 tonnes, but processing capacity may double if planned units come on line in the next few years.

SMALL-SCALE AGAR PRODUCTION

Small-scale, village level agar extraction has been tried in Thailand, Malaysia and India. In India, an attempt was made to adapt the Thai method to the village situation in Tamil

Nadu, India. Inadequate dehydration before sun-drying has been difficult to overcome. In the village setting, the source of heat energy may also present difficulties. Wood is often the only fuel available and is costly.

Economic models can be used to set production targets, and demonstrate what technical problems influence profitability and how. However, there is a paucity of economic data and analysis for various agarophyte production systems now in use or being tested in the region.

GEL STRENGTH

International trade in dried Gruciluriu is based largely on gel strength; the higher the gel strength, the higher the price paid. Its measurement raised some questions: although the processing industry recognizes a standard method, it is not always used by investigators. Unlike other countries, the Indian industry does not use chemical treatments to upgrade gel strength. It is believed that Indian Gracilaria does not respond to such treatment. Local strains may be too high in sulfate or there could be some other environmental influences. Poor post-harvest handling could play a role.

There was disagreement on whether or not village industry could produce an agar of sufficient quality for the market. Several industrialists felt that it cannot and that villagers are better off concentrating on seaweed farming. Adaptive research could address the major problems in the village context - filtration, water removal and sanitation. Perhaps improved technology at this level is available from other countries. Japan has had a village industry for centuries. There are many small-scale agar plants in Taiwan, whose experience can be applied to local conditions in the BOBP countries.

All BOBP countries import agar, particularly the higher grades. The consensus is that world demand is increasing, but there is a lack of detailed information on the trade. A comprehensive study of world demand is needed in order to guide farmers, processors, investors and government policy makers.

MARKETING

Although trade statistics are often questionable, world production is somewhere between 7,000 and 10,000 tonnes annually. Of this, about 3,700 tonnes enter international trade. Dried seaweed (Ceylon moss) has been exported from Sri Lanka since at least the mid-19th century, according to historical data.

The present day trade is only a fraction of past production: 5 tonnes compared to 150 tonnes more than 100 years ago! The decline does not appear to be due to the resource base, but rather due to changes in the fishery. Seaweed collection is not competitive with shrimp fishing as a source of livelihood.

In Sri Lanka there is an interesting local market for dried Gracilaria sold in small packets for home production of agar. Porridge flavoured with coconut milk and lemon juice is popular among fishermen in the seaweed-producing areas. It is sold by the glass in small roadside stalls.

Home-made jellies are also sold. Malaysia is one of the region's major consumers of agar and agar products, amounting to M$ 6 million annually. The Malaysian market could have a strong impact on future developments in farming and processing in the region.

MANAGEMENT OF GRACILARIU RESOURCE

There is considerable potential for increased production from well managed natural stocks; experiences in Chile are an outstanding example. Production there increased from 80 to 600 tonnes after management measures were introduced.

However, an effective management program depends upon knowledge of the reproduction, growth cycle, growth rates, regeneration capacities, productivity and the influence of environmental factors on biomass production of the stock in question. Unfortunately, very little work has been done on natural stocks of agarophytes in the Bay of Bengal region.

Stocks are reported to be over-exploited in the main producing area of India, though firm evidence is lacking because there have been no recent surveys of the resource.

Agarophytes have been inventoried at 31 locations in Indonesia, mainly in Java, Sulawesi, and Bali. In Thailand there is a potential harvest of Polycavernosa (closely related to Gracilaria) from the large number of fish cages in southern Thailand. During the discussions, it was suggested that this might lead to new farming systems utilizing natural spore collection on discarded fish net.

The confused state of taxonomy was a recurring topic for resource management, species must be clearly known before individual stocks can be identified. Comparative studies of species could also be advantageous for culturists as well as resource managers.

3

Status and Culture of Tropical Agarophytes

INTRODUCTION

The increasing demand for agar, agarose and agaropectin is primarily due to the development of new applications of these seaweed products in the food industry, in specialized laboratory methods such as immunodiffusion, diffusion and chromatography techniques, bio-engineering, microbiology and biochemistry. A stable source of raw materials for the manufacture of these seaweed products is essential to meet this increasing demand.

At present, there are two major sources of tropical species of agar-producing seaweeds, namely from gathering natural stocks and from farming/culturing of these species. Production through gathering of natural stocks depends on the availability of harvestable stocks; this is greatly influenced by seasonal changes in the weather (monsoons) which in turn make supply seasonal. In addition, production is highly influenced by the harvest pressure of the previous season. These factors lead to a tendency for production from natural stocks to be unreliable. Production through culturing is more predictable and stable, and targeted outputs are easily attained. Cultivation entails man's intervention in the process of production. It involves a wide range of techniques, from the simple management of natural stocks to the complex propagation of selected clones and genetically bred hatchery seedstocks in capital intensive situations (Neisch 1978).

CONSIDERATIONS IN THE SELECTION OF SPECIES AND SITES FOR CULTURE

The following points are important when considering the selection of species to be used and sites for culture:

(a) The species targeted for culture must be fast growing and capable of being propagated utilizing vegetative propagules (cuttings) or spores. Species with large and robust thalli are preferred because they can produce a large amount of biomass within relatively short cropping periods.

(b) It must have a high content of good quality agar. These considerations require that the available species be screened. Comparative studies on their productivity should therefore be conducted to determine the species to be selected. Because the properties of agar (gel strength. gelling and melting temperatures, amount of sulfate groups) differ among species, it is essential that the correct name be applied. Thus the taxonomy of the species should be clarified. The names are indices to the type and quality of agar that the Gracilaria contain and are used as the basis for determining prices.

(c) The selection of sites is a very important factor to consider in both pond and field culture, because this will ultimately determine the success or failure of the farming venture. In general, sites which support natural stocks of the species to be cultured are good sites. If culture is planned in an area where no stock of the chosen species exists, then one should select sites which have ecological conditions comparable to the site where the stocks are found. This requires the gathering of data on parameters such as salinity ranges, nutrient (i.e. nitrate and phosphate} levels, turbidity, and type of substrate. Other ecological parameters such as degree of exposure to waves, depth of the water and accessibility should also be included in the evaluation of sites.

Natural Gracilaria stocks are generally found in areas which are characterized by calm water. high nutrient levels and shallow areas with sandy-muddy substrates. Sites for field culture should be located in areas which are protected from waves since Gracilaria have fleshy and fragile thalli and are therefore easily removed or broken by wave action. Species which are found in commercial quantities generally grow quite well in slightly brackish and highly fertilized waters, thus protected bays and lagoons are preferred. In tropical areas, Gracilaria species with a high potential for culturing are also found on reef, sandy flats, sandy, rocky and wave-exposed areas.

PRODUCTION OF SEEDS

The term "seeds" as used here refers to both the vegetative propagules (cuttings) and spores which can he utilized as planting materials. The importance of the availability of a local source of seed materials is emphasized. These seedstocks may come from natural stocks or from cultivation. Gracilaria exhibits triphasic alternation of generations consisting of the gametophyte, sporophyte and carposporophyte. The latter, however, is microscopic and is parasitic on the female gametophyte.

Thus, the large somatic stage consists of the gametophyte and the sporophyte. The results of studies on the populations of these two generations have shown that the sporophytes are generally more dominant than the gametophytes, i.e. they are usually larger in size and the population makes up the bulk of the biomass. It has also been shown that under pond culture conditions the alga may not develop reproductive stages. These factors are relevant in the selection of the type of "seeds" to be used in culture.

Vegetative Propagules (cuttings)

Healthy stocks are selected for this purpose. They should have thalli which are fleshy and elastic in texture, dark reddish brown in colour, and are robust, well-branched with smooth and shiny surfaces. In addition, they must be clean, i.e. free from dirt and epiphytes. The preparation of the "seeds" varies according to the type of culture used, and these are described in appropriate sections of this paper.

Spores us Seeding Material

Two methods are presently used in the production of sporelings from spores: natural spore-recruitment and induced spore-seeding in hatcheries.

Natural Spore Recruitment

The use of spores as seeding materials for the production of sporelings for culture requires the availability of natural stocks. In the natural spare recruitment method artificial substrates such as ropes, rocks and netting are used. Ropes and netting are generally preferred because of their ease of handling during both the process of recruitment and cultivation. The use of ropes is a proven technique in the West Indies. The ropes are anchored or tied to wooden stakes among dense populations of Gracilaria. They are left in the area for about two weeks to allow the naturally shed spores to settle on them. The sporelings developing from the spores become visible after three to four weeks, and the seeded ropes are then transferred to the culture sites for outgrowing.

Hatchery Production of Sporelings from Spores

The production of sporelings from spores requires skill in recognizing fertile materials, and selecting them from available stocks. Recognition of the fertile female materials is quite easy because of the highly recognizeable presence of cystocraps. These fertile structures appear as elevated mammilate "bumps" with dark coloured contents on the surface of the thalli. Cystocarps which have already shed their spores are pale in colour. In contrast the recognition of fertile sporophytic thalli is quite difficult because the tetrasporangia are microscopic and do not form apparent recognizeable structures. The use of a stereomicroscope is quite indispensable. Under low magnification, the tetrasporangia appear as dark purplish microscopic spots on the surface of the thalli. The presence of these structures has, however, to be counter-checked by examining sections of the suspected thalli under a compound microscope. The tetrasporangia are of the cruciate type and are embedded in the cortex of the thallis.

The hatchery production of sporelings requires a land-based set-up which may be sited some distance away from the outgrowing area. The set-up was first used in Penang, Malaysia. It is quite similar to those used in the seeding facilities for Porphyra and Laminaria although less sophisticated. It consists of a seeding tank with provision for control of water depth and an adjustable seeding material support structure.

This structure consists of a wooden frame strung with monofilament netting material which fits into the seeding tank. It is provided with a mechanism that allows its height from the bottom of the tank to be adjusted.

Various types of substrates, e.g., pieces of gravel, shells or lines can be seeded. These materials are placed at the bottom of the seeding tank. The use of lines as a setting substrate requires an additional structure, i.e. a frame which fits the tank when laid flat on the bottom.

The line may be monofilament nylon (90 lbs test as used in Malaysia), nylon braided ropes (ca. 3mm dia) or plastic tying materials. The line is wound evenly around the seeding frame, which is then placed at the bottom of the tank.

Fertile Gracilaria material collected from natural stocks is brought to the seeding facility. The moist materials are placed in containers, such as Styrofoam boxes provided with aeration holes, to avoid stressing which can affect the shedding and/or viability of the spores.

The fertile Gracilaria are then placed on the seeding support structure submerged in water in the seeding tank. The distance of the seeding materials from the seeding substrates (rocks, stones or lines in frames) should be adjusted to ensure the uniform distribution and density of the spores.

The seeding materials are left in the tanks for two to three days to allow shedding of spores. The seeded materials may be left in the tank for another one to three days to allow the spores to germinate and become completely attached to the substrates. The seeded rope is transferred to a holding tank or to a nursery area while waiting for the spores to develop into sporelings.

PRODUCTION AND SELECTION OF HIGHLY PRODUCTIVE SEEDSTOCKS

The utilization of highly productive species/strains with high quality natural product is essential for the successful farming of Gracilaria, or any crop for that matter. In species where it is easy to manipulate the individuals to be improved, hybridization utilizing the gametes may be the normal and logical way of improving the seedstocks. However, in Gracilaria and other species, where the techniques for manipulating the sexual process to produce hybrids are not well known, different methods of selection may be used.

One of these methods is species/strain selection. Different species/strains present in a natural population e.g., different species, various coloured strains, morphologically different thalli are selected. These different species/strains are then multiplied through cloning. Comparative ecophysiological studies are then conducted using these stocks. In situ studies on seasonality and production capacities of the various species/strains relative to changes in ecological factors such as salinity, temperature, nutrients and water movement are important in defining the field conditions which determine the productivity of the various strains, as well as the seasonal aspects of reproductive states and capacities of the various forms.

Physiological studies to determine the range in tolerance of the various species/strains to changes in ecological factors can be determined by manometric techniques using a respirometer, where the various factors can be accurately controlled. The photosynthetic/respiratory responses can be monitored under varying conditions of stress. Analysis of the content and quality of agar in the various strains is also an indispensable tool for determining the qualities of the various strains or species as a basis for the selection of seedstocks.

METHODS OF CULTURE

Pond Culture

Although several species of agarophytes belonging to the genera Gelidium, Pterocladia and G r a c i l a r i a have been reported to he produced commercially through some form of farming in several countries (such as Japan, China, Republic of Korea, Vietnam, India and the Philippines), it is in Taiwan where the production of Gracilaria through pond culture has achieved a high degree of success. Here, an annual average of 12,000 tonnes of dried Gracilaria has been produced in recent years (Chiang 1981).

As stated earlier, the genus Gracilaria is characterized by the alternation of three somatic generations. Of these, it is the macroscopic, gametophytic and sporophytic stages which are used as planting materials in pond culture. Although the reproductive potential of Gracilaria through spores is high, vegetative propagation by cuttings is used in pond culture at present because of the very high regenerative capacity of the plant and the simplicity of the method. However, hatcheryproduced seedlings from spores have been demonstrated to be superior in the open field culture of Gracilaria (Doty 1986).

Of the several species of Grucilaria presently used (e.g. chordu, G. tenuistipitutu, G. edulis, G. verrucosa, G. lichenoides, G. compressa and G. giga), G. verrucosa is the most popular due to its ability to adapt to a wide range of ecological conditions in ponds, its higher production rates and better gel quality. The culture of Gracilaria started in 1962 in southwestern Taiwan. Production in ponds is primarily influenced by three ecological factors, namely salinity, light and temperature. High production is recorded during the warmer months and growth is slow during winter. High light intensity adversely affects growth, and light conditions are controlled by adjusting the water depth in the ponds. Salinity of 15 to 24 ppt appears to be optimal for growth. The increase in salinity during the summer months is controlled by the addition of freshwater, thus farms need to be located near freshwater resources.

Site Selection

Successful pond culture of Gracilaria depends greatly on the selection of appropriate sites. The following criteria are recommended for the selection of pond culture sites.

- the site should be located near seawater and freshwater resources;
- the area should be protected from strong winds;
- the pond bottom should be at or near the zero tide level and have a sandy-muddy substrate;
- the pH of the water should be slightly alkaline (i.e. 8.2 to 8.7).

Gracilaria are euryhaline and can grow in brackishwater under a wide range of salinity, although 15 to 24 ppt has been found to be optimal. Rises in salinity during the sunny months due to evaporation losses (reaching values as high as 35 ppt), or falls (to as low as 8 ppt) during the rainy season are detrimental to the crop. The maintenance of optimal salinity in the ponds requires a readily available supply of freshwater and seawater. The ponds should be located in areas protected from strong winds because there is a tendency for Gracilaria to accumulate on the leeward side of the pond. The formation of thick heaps of Gracilaria in one side of the pond causes shading, which has adverse effects on growth.

Water management is greatly influenced by the tidal changes in relation to the elevation of the pond bottom. Ponds located in areas where the bottom is at, or a little above, the zero tide level are easily managed as water exchange is easy.

Culture Ponds

The average size of ponds for the culture of Gracilaria is about one hectare or smaller. Smaller ponds are easier to manage than larger ones because they are less susceptible to the influence of winds. Pond management is also easier when Gracilaria is polycultured with shrimp and/or crab. Provision of entrance and exit gates facilitate proper water management.

The depth of the ponds vary from 50 to 80 cm. The bottom is generally of clayish loam, silty loam or sandy loam. It has been observed that in ponds with a sandy bottom, Gracilaria easily gets buried due to the effect of wind. This problem, however, could be resolved by increasing the depth of the water during windy periods. In larger ponds, wind-breaks consisting of bamboo slots are installed perpendicular to the direction of the wind to prevent the seaweed being transported to one side of the pond.

Culture Method

The following method is generally followed in the pond culture of Gracilaria. The ponds are dried for several days, and then water is introduced. Healthy stocks (see earlier criteria) are selected as planting materials. The planting material is transported from its source to the pond site early in the morning to prevent its exposure to the sun. During long-distance transport it is frequently sprinkled with seawater, and perforated bamboo or plastic pipes are inserted into the bottom of the heap to provide aeration.

The plants must be placed in the water of the pond immediately upon arrival. The planting material is then cut into pieces and is broadcast uniformly on the bottom of the pond. In Taiwan, stocking is usually carried out in April at a density of 5,000 to 6,000 kg of chopped Gracilaria per hectare.

Pond Management

The water is maintained at a depth of approximately 30 to 40 cm above the algae. However, the depth is increased to 60 to 80 cm during the warm summer months to prevent a significant rise in the water temperature. Water depth is also increased during the cold winter months to avoid temperature drops below 8°C which are lethal to Gracilaria.

Frequent exchange of water is necessary to maintain the optimum temperature in the ponds. The water is changed every two to three days, when about 50 to 75% of it is drained and replaced with fresh seawater.

The growth of Gracilaria is enhanced by adding either organic or inorganic fertilizers. In Taiwan, weekly application of urea at 3 kg per hectare was found sufficient. Fermented pig manure may be applied at 160 to 180 kg per hectare two to three days after changing the water.

Harvest and Post-harvest Activities

Under optimum conditions, the crop may be harvested two to three months after seeding. Cropping may be done every 10 to 40 days, either manually or by using scoop nets. The frequency of harvests is dictated primarily by the market price and the season. Approximately 30 to 40% of the biomass is harvested during each cropping. The crop is thoroughly washed in pond water to remove the silt, sand, pieces of shells and other extraneous materials such as snails and other algae. The clean Gracilaria is spread uniformly on bamboo screens or plastic sheets for drying. An average wet to dry ratio of 7:1 is generally attained.

Standards set by the Bureau of Standards in Taiwan for the export of dried Gracilaria require that the product should not contain more than 1% of mud and sand, not more than 1% shells and not more than 18% other seaweed species. Moisture content should not exceed 20%. Dried Gracilaria is then packed into sacks of 100 kg weight which are either exported or sold to local processing plants. Ten to twelve metric tonnes of dried Gracilaria are produced in a hectare of pond.

Polyculture with Shrimp and/or Crab

Polyculture with shrimp (Penaeus monodon) and/or crab (Scylla serrata) is carried out mainly in Ping-tung prefecture in southwestern Taiwan. Stocking material for a hectare of farm consists of 4,000 to 5,000 kg of Gracilaria, 5,000 to 10,000 crabs and 10,000 to 20,000 shrimps. Crushed trash fish and snails are generally used as feed for crab. Crabs are harvested after three months, and the shrimp after four to seven months. Survival rates as high as 80% for crabs and 80 to 90% for shrimp have been documented, making this polyculture one of the most profitable aquaculture methods in Taiwan. The net income from polyculture has been shown to be three times that from monoculture.

In the Philippines a group of 40 fish farmers in Iloilo are producing Gracilaria in their multi-crop fish farms, using cuttings as seedstocks. Unlike Taiwan, they do not use fertilizer;

instead, they allow daily water exchange by tidal flow to ensure adequate nutrient supply for the seaweed. The ease of water management is enhanced by siting the bottom of the ponds at, or slightly above, the zero tide level. Problems with water management are encountered when ponds are located away from both fresh water and sea water sources, and where the level of the bottom of the ponds is way above the zero tide level making water changes a major problem. In such ponds water changes can only be facilitated during extreme high tides, and maintenance of adequate nutrient levels and control of water temperature are difficult. Large size ponds are subject to problems caused by wind generated waves referred to earlier. Provision of wind breakers such as "bamboo blocks" and "net blocks" in the ponds perpendicular to the direction of the prevailing winds (as practiced in Taiwan) has proved effective and led to increased production compared to ponds without them.

The development of blooms of other algae such as filamentous blue greens and greens is a major problem in the Philippines. These algae may totally displace Gracilaria, or else they become so mixed with Gracilaria that it is impossible to remove them, resulting in dried materials of very low quality. The problems of blooms of epiphytes and other weeds are being controlled by the introduction of grazers such as Tilapia and milkfish. However, the size and number of these grazers must be controlled otherwise these may consume the Gracilaria. Pond culture of Gracilaria is also being carried out in Vietnam, Thailand, Indonesia, Hainan and Hawaii.

Field Culture

Several methods are currently being utilized in the field culture of Gracilaria. The "seeds" used in these various methods are cuttings or sporelings produced from spores.

Fixed Bottom Type of Culture

Long Line Method: In this method, three-strand polyethylene ropes 5-8 mm thick and five metres long are generally used. Other materials such as coir or abaca ropes may also be used. In the West Indies, where the mariculture of sea moss (Gracilaria) is most successful, both sporelings produced from spores and cuttings are utilized as "seeds" in the fixed long line method as well as in the raft method. The seeding of long lines using cuttings consists of untwisting the rope and inserting bunches of cuttings between the strands of the rope, the seedlings passing twice or three times through the rope. This method ensures that the cuttings are securely held in place. The seeding of the ropes is done in the shade, where the seedlings are placed in basins of sea water to prevent them becoming dehydrated.

The support system for the long lines consists of rows of wooden stakes driven into the ground. The distance between stakes in the row is about one metre, while the distance between rows varies between 4-5 metres or longer. The seeded rope is tied to a stake, stretched tightly and rhe other end tied to the opposite stake in the next row. In the West Indies, two seeded ropes are usually tied to a pair of stakes.

The same technique is supplied for ropes seeded with spores. In Malaysia, unbraided plastic rafia seeded with hatchery produced sporelings was used as long lines. In India, longer ropes are used and additional stakes are provided to prevent the ropes from sagging to the ground. The distance of the seeded ropes from the ground varies depending on the depth and clarity of the water as affected by tidal changes. This'method of field culture is presently being utilized in the West Indies, Myanmar, India, Brazil and Sri Lanka.

Chilean Method: The field culture method presently used in Chile is quite simple but has proved very successful. The anchoring system for the seedstocks (cuttings) consists of plastic bags one metre long, 0.1 mm thick and 4 cm in diameter. The plastic bag is filled with sand and the two ends are knotted. Five bunches of seedstock are fixed to the same side of the sand-filled bags by rubber bands. The seeded bags are arranged in rows on the substrate and are positioned in such a way that the bag serves as an anchoring weight. The middle of the cuttings are weighted by the bag against the surface of the substrate and their tips freed.

The use of rocks as anchoring units for the cuttings is effective. However the use of both sand-filled bag and rocks as anchoring materials for cuttings was only efficient in sites which are not exposed to strong surges or waves. Evaluation of the efficacy of this system showed that more than 60% of the biomass could be lost due to removal of thalli by strong surges, and rocks were observed to be more easily dislodged from the substrate than the sand filled bags.

Improvement of Substrates: This method can only be utilized in areas where natural beds are found. The introduction of artificial substrates such as rocks, shells, bamboo or wooden stakes, etc. in the site provides additional favourable substrates for the spores to settle or fragments of thalli to attach. Most of the beds of commercially important species of Gracilaria are found in shallow bays and coveswhere the substrate is generally particulate (sand-muddy). The survival and growth of spores is greatly enhanced by the availability of solid substrates. Studies of the influence of solid substrates on biomass production have shown that production more than doubled in portions of the beds where solid substrates had been introduced.

Raft Method

The raft method is popularly used in West Indies in areas where bamboos are available. The raft generally measures 2 x 4 metres in size. Seeded long lines are tied to the shorter bamboo frame, stretched tightly and fixed to the opposite side of the raft. The distance between ropes varies, but about 12 to 14 lines may be planted to one raft.

In India, coir ropes fabricated into network of 7 cm mesh size was used on a 2 x 2 metre raft. The same method of seeding was utilized.

A variation on this method has been developed in the People's Republic of China. Pieces of rope about 70 cm long seeded with Gracilaria cuttings are hung horizontally on the low fixed raft or vertically along a long line floating raft.

4

STATUS AND CULTURE OF AGAROPHYTES IN CHINA

INTRODUCTION

A number of agarophytes, including Grarilaria, Gelidium and others, have been used in China as marine vegetables or traditional medicinal herbs for over 1,000 years. However, these seaweeds became an industrially important raw material for processing agar only some decades ago. Agarophytes in China are mainly represented in three genera of Rhodophyta, Gracilaria, Porphyra and Gelidium. Among more than 30 species of agarophytes, Gracilaria verrucosa, G. tenuistipitata, Gelidium amansil and Porphyra haitanensis are the major ones. In addition, other red seaweeds, such as Pterocladiu sp., Gelidiella sp. and Ceraminum sp. are often collected and mixed with the main agarophytes mentioned above and used for producing agar.

China is a major world producer of cultured seaweeds. The total annual production of cultured seaweeds is about 300.000 dry tonnes** Of these, Laminaria and Porphyra account for 200,000- 250,000 tonnes and 20,000 tonnes respectively. Gracilaria production is about 5,000 tonnes per annum (Shang, 1976; Chiang, 1981, Yang et al., 1981; Chueh and Chen, 1982; Zheng, 1987).

In China, Laminaria is a main marine vegetable and raw material for extracting iodine, mannitol and alginate. Porphyra is also a popular marine vegetable. But in recent years some phycologists have observed that a considerable quantity of Porphyra thalli obtained from the later harvests during the season can be used for processing agar. These later croppings have less edible value. At present Gracilaria and Porphyra are the most important raw materials for producing agar.

Although Gelidium has long been considered as the best raw material for processing agar,it is not the ideal cultured species because of its slow growth and low yield. On the other hand, Gracilaria is easier to farm. It is expected that a higher output can be obtained in farming conditions. Shang (1976) reported that the average productivity of dry Gracilariu in Taiwan ponds was 7-12 t/ha/year. A few years later Chiang (1981, P. 573) observed that 16-43 t/ha/year could be expected. So it is believed that there will be a substantial development of Gracilaria culture in China to meet the increasing demand of the agar market.

In the present review I shall first discuss some general ecological characteristics of Gracilaria which are especially important to its commercial production, and subsequently review the farming practices developed for the species.

BIOLOGY OF GRACILARIA

Gracilaria verrucosa is the most common species in China. It grows fast and has a high agar content. Following is a discussion on the influence of the various environmental factors on the release and germination of spores and the growth of thalli of the spores.

Influence of Environmental Factors on Spore Release

Both carpospores and tetraspores are released naturally into seawater after their maturation. Experiments showed that the maximum quantity of spores is released from 8 a.m. to 10 a.m., gradually decreasing thereafter. The minimum quantity is released between 10 p.m. and 6 a.m., after which the next maximum will take place. Besides the diurnal changes in release rhythm, the quantity of spores released also depends on ambient environmental factors. (Zheng, 1987).

Deciccarion

In general, when a mature Gracilaria is taken out of seawater and kept in a shady spot for two to four hours (air temperature 15-25°C), the tetraspores or the carpospores will be released if the plant is dipped in seawater again.

Seawater Temperature

Experiments have shown that the highest number of spores are released at 20-25°C. According to field surveys, the reproductive season of Gracilaria is between June and August in northern coastal provinces or between March and May in southern China. During these seasons, the natural seawater temperature is 20-25°C. The experimental results coincided with natural reproductive seasons (Zheng, 1987).

Specific Gravity of Seawater

When mature plants were kept in seawater of different specific gravities, those in seawater of lower specific gravity would release spores earlier than those kept in water of higher specific gravity. If the plants are put in seawater with an obviously low specific gravity(<1.005), the released spores will swell and even break up due to osmosis caused by the low salinity and most of them will not complete normal germination. In order to obtain good results in the collection of spores, the optimum specific gravity range of 1.015 (19.0-20.2 ppt) to 1.020 (25.5-26.9 ppt) should be observed (Zheng, 1987).

Influence of Environmental Factors on Germination and Growth of the Spores

Gracilaria spores just released into seawater measure around 30 um in diameter but this varies with species. Tetraspores are slightly bigger than carpospores. For example, the tetraspores of G. tenuistipiruta have a diameter of 24-55 um, and the carpospores 23-40 um. Soon after release, the spores will attach to substrates and start first cleavage of cells. The germination and growth of spores are also influenced by amhiect environmental conditions such as seawater temperature, light intensity and salinity (Zheng, 1987).

Seawater Temperature

To examine the germination and the growth of the spores, specimens attached to slides for 12 hours were placed in sterilized seawater enriched with 1 ml of 1 M KNO, and 1 ml of 0.1 M KH2PO4 per litre of seawater medium, and incubated with a light intensity of 400 Lux, photoperiod of 10:14 for different periods of time (Zheng, 1989).

The results showed that both tetraspore discs and carpospore discs need a seawater temperature higher than 15°C. Even if all other environmental conditions are suitable for their growth, they would survive but grow very slowly (Zheng, 1987).

Light Intensity

Light intensity is one of the most important factors that influence the germination and growth of spores. When the light intensity is stronger, the rate of germination and growth of spores are higher within 3,000 Lux. If the spores attached to slides are kept in a dark place they will die in 20 days.

Specific Gravity of Seawater

Although Graciluria plants prefer to inhabit estuarine areas, their spores are unable to withstand seawater with low salinity. If the spores attached to substrate are put in seawater with a specific gravity below 1.010, (12.5-13.7 ppt) their cells swell up due to the absorption of water into the cells. The colour of the pigments in the cells becomes pale, and the spores eventually die. Experiments proved that the optimum specific gravity of seawater for germination and growth of spores ranges from 1.018 (23.0-24.2 ppt) to 1.025 (25.5-26.9 ppt) (Zheng, 1987).

The Effect of Environmental Factors on Growth of Gracilaria Thalli

The effects of environmental factors on Graciluria thalli are similar to those on their spores, but not entirely the same. For instance, as mentioned above, spores kept in seawater with a specific gravity below 1.010, will break up and die, but the thalli can grow very well in the same conditions. In view of this, ambient environmental factors required by Gracilaria thalli must be studied.

Specific Gravity of Seawater

Gracilaria are euryhaline seaweeds. Under natural conditions, the specific gravity in which the plant can grow out ranges from 1.005 to 1.026 (5.2-38.1 ppt), Experiments and field surveys have shown that the optimum specific gravity is 1.010-1.020 (11.3-30.1 ppt) where freshwater regularly flows in (Zheng, 1987).

Seawater Temperature

Gracilaria are eurythermal plants which can grow at 5-30°C. Optimum temperature varies with species. For example, the optimum temperature for G. verrucosa is 15-25°C. They can be found in northern China during May to August, but in south China only in winter and spring. In summer, the growth of Gracilaria is almost completely stopped in the south until late autumn when the water temperature drops below 25°C and the seaweeds resuscitate.

Field surveys have shown that the optimum seawater tzmperature for G. tenuistipitata, a species mainly distributed in Guangdong and Hainan Island, is identical with G. verrucosa. When the water temperature is 30° C, its diurnal growth rate measures 0.1-0.2 cm/day but as the temperature continues to fall to 28°C, the growth rate increases to 0.4-4.5 cm/day. When the temperature is from 15-25°C, growth rate can be higher than 1 cm/day.

Light Intensity

The growth of Gracilaria requires a high light intensity. It has been shown that growth rates of seaweeds vary when they are planted at different depths of water (Zheng, 1987). The study showed that the Gracilaria can grow well in water less than 1 m deep where the seawater transparency is around three metres. If the seaweeds are planted in seawater deeper than three metres, they not only stop growing but start rotting as well.

Dissolved oxygen was measured in seawater collected from different water depths where Gracilaria is located. The results showed that Graciluria exposed to the maximum light intensity also has more intense photosynthetic activity. As the plants are located beneath 3 metres depth, their respiratory intensity exceeds the photosynthetic intensity. This explains the finding that the amount of dissolved oxygen in the seawater of the experimental groups was less than that of the control group in which no Gracilaria were present.

FARMING

China has more than 30 years' history of Gracilaria farming. According to the specific biological characteristics of the seaweeds, different methods of collecting spores and farming thalli have been studied and practised. This includes collecting natural spores, artificial outdoor collection of spores, indoor spore collection, mud flat culture, floating raft culture and pond culture.

Rearing of Seedlings

Outdoor Collection of Spores

The procedure involves four steps: selection of site. preparation of substrate. collection of parent plants (tetrosporophyte and/or carposporophyte) and collection of spores.

Site Selection: A suitable site for collection of spores should be an intertidal flat area with a hard bottom and clear seawater with a specific gravity range of 1.010-1.025 (12.5-33.4 ppt). A large spherical dwarf dam 20-30 cm high is made at the selected site to store seawater after ebb tide.

Preparation of Substrate: Cheap substrates like stones, shells or broken corals can be used. These substrates have a clean surface for easy attachment of spores. If the substrate is stone, each block should weigh about 0.5 kg, the total amount is about 600 tonne stones per ha. For shell, 120 tonnes per ha is sufficient. The substrate is spread on the selected site. In recent years, artificial fibre (e.g. nylon, polyethylene) ropes or nets have been used to collect spores. The texture of these materials is good but the cost is higher. Used or worn fishing nets can be used as cheaper substrates.

Collection of Parent Plants: The parent plants must be fully mature with a dull red-brown colour and without rotted spots. In fully mature plants, the sporangia are easily seen on the entire surface of the plant body.

Treatment of Parent Plants: Parent plants are washed with clean seawater to get rid of mud or other matter and allowed to air dry on bamboo curtains for two to four hours until irregular corrugations appear on the plant surface. At this time, the seaweeds can be spread in a selected site with substrate. The quantity of parent plant required is about 300 kg per ha. As soon as the parent Gracilaria are planted in seawater, they release a great quantity of spores into the seawater. The spores will settle down on the substrate in one day. Under natural conditions, the germination of spores is very fast. They will turn into dish-shaped thalli in five days. Another method is to put the desiccated seaweeds into a vat and stir them vigorously with a stick to stimulate release of spores into seawater. About two hours later, the seawater containing a huge quantity of spores can he sprayed into a selected site.

Indoor Collection of Spores

The main steps followed in indoor collection of spores are similar to those in outdoor collection. The only difference is that the spores attached to the substrate will be kept in indoor tanks for rearing until they grow into young plants.

Culture Methods

Mud Flat or Intertidal Culture

The main feature of this culture method is that the spore collection and farm sites are usually the same. In southern China, in late autumn or early winter, young Gracilaria can grow up to S-6 cm long as they enter a period of faster growth rate. At this stage, the substrate to which young Gracilaria are attached are kept in lines on the bottom, which serve as a walkway, at an interval of 30-40 cm. Routine management practice involves removal of miscellaneous seaweeds, collection of herbivorous gastropods and so on. Women and children often do this. In a normal season, 1,500 kg (dry) of Gracilaria can be harvested.

Floating Culture

The method has been adapted from kelp farming. In a suitable season, such as January in southern China or May in the northern part, young seaweeds collected from other fields or reared in indoor tanks are pulled up from the substrate and inserted into "seedling ropes", which are made of palm thread or artificial fibre. In each 20 metre rope, 200 pieces of seaweed can be inserted at 10 cm intervals. The seedling ropes are then fixed to a floating raft. Three months later, the seaweeds will reach a length of over one metre and can be harvested giving yields as high as 3000 kg (dry) per ha.

Pond Culture

Pond culture of seaweeds has been adopted by Taiwan's phycologists and farmers since the 1960s. Before 1962 agarophtyes were very scarce in Taiwan, and had to be imported. Phycologists and farmers tried to rear Gracilaria in ponds with milkfish. Unexpectedly good results were obtained and many farms began to change the main cash crop from milkfish to Gracilaria.

At present, Taiwan produces 12,000 tonnes fresh weight of Gracilaria annually from 300 ha of ponds (Chiang, 1981). The stock seaweeds are cut into pieces and spread in the fish pond at a density of 5-6 tonnes of fresh thalli per ha. Chemical fertilizer or pig manure is applied regularly. Every 30-45 days, most parts of the seaweeds can be harvested leaving the rest of the body in the ponds as stock. The harvesting period lasts six months from June to November in the northern region. This method is also being used in the southeast region where the lesser inputs and higher production make it an attractive proposition.

GENERAL COMMENTS

Although satisfactory results have been obtained by several Gracilaria culture facilities, there have not been the great advances in production to match those achieved in Laminaria or Porphyra farming. The main reason for this is a lack of balance between production cost and market price. Indeed the lower market price is inhibiting the development of Gracilaria farming. In recent years, Gracilaria farming has become popular because of the increasing demand from the agar processing industry. A main production zone of Gruciluria has grown up in Southern China (including Guangdong, Guangxi and Fujian provinces) where seawater is warmer and rich in nutrients suitable for growing Gracilaria plants. Liu (1987) reported that there are four species (G. verrucosa, G. tenuistipitata, G. gigas and G. bursa-postoris) which are suitable for farming. His experiments and investigations have shown that it is possible to obtain an average yield of two to three tonnes per ha. He suggested that the acreage under farming of Gracilaria can be expanded to 6700 ha in Guangdong Province to produce 15,000 tonnes (dry) of high quality Gracilaria. Technical progress and advances in Gracilaria farming will further reduce the production cost and improve product quality. The following suggestions are made to improve the culture, production and utilization of this agarophyte:

(1) High priority should be given to selecting species with faster growth rates and higher contents of good quality agar.

(2) Adequate measures should be taken to protect natural stocks of Gracilaria to prevent overexploitation of natural stock.

(3) Polyculture of finfish and shellfish with Gracilaria offers several advantages. For example, the autophyte seaweeds can improve the quality of water contaminated by cultured animals; Gracilaria is a superior fodder for abalone which has a higher price in the world market. Gracilaria farming can supply feed to abalone farms, while abalone farming would also help to promote agarophyte farming.

5

Status and Culture of Agarophytes in Indonesia

INTRODUCTION

Studies on the biology and agronomy of Eucheuma were conducted at Geger Beach, Nusa Dua, Bali, Indonesia from June, 1986 to October, 1989. The experimental site is also the farm of about 50 families who are cultivating Eucheuma coftonii and Eucheuma spinosum. During the experimental period, Gracilaria lichenoides was observed growing among the Eucheuma plants on stakes, ropes, lines and rafts.

Since we were not interested in the culture of G. lichenoides, these seaweeds were considered as pests and were usually removed from near or around the Eucheuma during farm maintenance operations.

Recently (September-October, 1989), we became interested in the G. lichenoides because we observed that the farmers were also gathering these plants and selling them to the local traders who, in turn, wash the dried Gracilaria in fresh water and after re-drying sell them at the market at Denpasar, Bali.

Preliminary cultivation studies on G. lichenoides Wild G. lichenoides were collected from the stakes, ropes, and rafts where they were growing. They were divided into uniform fifty gram cuttings and then tied with nylon nettings. The propagules (cuttings) were cultured in the same way as Eucheuma, by tying them to the lines attached to a raft (floating culture system) at intervals of 20,25, and 30 cm. along the line. The plants were harvested six weeks after planting just like the Eucheuma.

The growth ratio (harvest weight/propagule weight, H/P) was computed by dividing the mean harvest weight by 50 grams which was the weight of the initial propagule.

Distance of Planting H/P ratio

20 212/50 = 4.24

25 229/50 = 4.58

30 230/50 = 4.60

The data show that the Gracilaria can tolerate closer planting, and suggest that they can be planted alternately between Eucheuma plants.

The percent dry matter yield (% DMY) was determined by drying 2070 grams of fresh G. lichenoides; this produced a commercially dry weight of 180 grams. Hence, the percent dry matter yield (% DMY) = 180/2070 x 100 = 8.69%.

The wet-to-dry ratio was therefore: 100/8.69 = 11.5

Gracilaria verrucosa is one of the aquatic plants which have been widely researched in the fisheries sector of Vietnam. Because of its wide distribution, fast growth, long annual culture time and capacity for harvesting several times a year, it was chosen as a species for culture in brackishwater coastal areas. Further, its culture is labour-intensive and generates substantial economic benefits.

Since 1963 Vietnamese scientists, and experts from the German Democratic Republic led by Prof. Dr. W. Brucker, have carried out surveys to determine the genus, species and distribution of seaweeds along the coast from Mong Cai district (Quang Ninh province) to Vinh Linh district (Quang Tri province). Surveys carried out by the Research Institute of Marine Products, Nha Trang Institute of Oceanography, Nha Trang Fisheries University and University of Hue gave additional information for all brackishwater areas along the coast of Vietnam. The results showed that there are 16 species of seaweed along the coast of Vietnam, of which Gracilaria verrucosa is predominant in terms of its distribution area, biomass and agar content and quality.

Research on biological characteristics, seed production techniques and intensive cultivation of G. verrucosa has been carried out continuously since 1963. In 1977, the Ministry of Fisheries approved a method for the bottom culture of Gracilaria verrucosa with a productivity of 10 tonnes/ha/year.

Between 1979 and 1988, some localities applied this method effectively, getting productivity-of 10 tonnes/ha/year, but the technology has yet to become stabilized and applied on a large-scale. However, in small ponds (200-1000 m2) at the Research Institute of Marine Products at Qui Kim, Hai Phong, productivity averaged 35 tonnes/ha/year.

The overall results showed that G. verrucosa is distributed widely in brackishwater lagoons in Vietnam, where the bottom is sandy mud, muddy sand or muddy. If the bottom is sandy, the growth is low; if the bottom is muddy the individual size of the organisms is reduced. In general, seaweeds can tolerate salinity from 5 to 35 ppt, and a water temperature of 38°C to 40°C. For G. verrucosa, salinity from 10 to 22 ppt, temperature from 20°C to 25°C, pH from 7.5 to 8.5 and water depth from 40 to 80 cm are the ideal conditions. Experiments on pulse fertilization showed that a compound of fertilizers consisting of 3 mg/1 $(NH_4)2SO_4$ + 3 mg/1 $NaNO_3$ + 0.6 mg/l $Na_2H_2PO_4$ was the most effective.

Pulse seed fertilization should last for only 4 hours, and be carried out in light rather than dark conditions. To limit the influence of foreign seaweeds on Gracilaria verrucosa, we should:

- improve pond conditions by ploughing the bottom and adding lime and organic fertilizer (0.2-0.5 kg/m^2); culture seaweed at an initial density of 500 g/m^2;
- intensify water supply/drainage for culture ponds.

Though the potential is great, the production of cultured seaweed in Vietnam is still low. Even in the best year to date (1987), production was only around 700 tonnes (dry weight), one third of which was harvested in natural habitats. The area available for seaweed culture in the whole (20) country in the next decade will be about 2000 ha of which Thua Thien Hue province nas the largest area upto 770 ha in the Tam Giang lagoon system.

Topographical and Ecological Conditions at Thuan an Seaweed Farm in the Tam Giang Lagoon System

The Tam Giang lagoon in Thua Thien Hue province is a continuous system of large bodies of water. Outside, the lagoon is surrounded by a long line of sand stretching from the mouth of the Tu Hien river to the estuary of the O Lau river, connected with the sea by the two estuaries of Thuan An and Tu Hien in the East. Inside the lagoon is the mainland of Phu Loc, Huong Phu and Huong Dien districts and Hue City. The total lagoon system is over 60 km in length. Due to the gently sloping bank the inshore part of the lagoon is shallow. In the dry season, the average water level is less than one metre. The depth in the middle of the lagoon is about 2 m but in some places up to 4-5 metres. The'speed of the river current is about 4.5 km/hr in the dry season, but in the rainy season it can reach 20 km/hr. The bottom structure is different: along the coast it is often sandy or muddy sand, in the middle of the lagoon it is often muddy or sandy mud. At the mouths of rivers one finds crude sand and hard terrain. The tide is semi-solar and irregular, and its amplitude is low (about 0.5-0.8 metres).

Thuan An seaweed culture lagoon is 2 km west of Thuan An estuary and is surrounded by dykes with gates and canals for regulating water.

In,1977-78, 10 ha of experimental ponds comprising eight small 0.5 ha ponds and one large (6 ha) pond were constructed.

In 1983-84, three additional ponds of 11 ha, 22 ha and 17 ha were constructed.

Another 50 ha were added in 1985-86, of which 20 ha were transferred to Phu Tan agricultural cooperative (Hue city). The remaining 30 ha were divided into 2 ponds of 15 ha each. The system of seaweed culture ponds in the Chuong lagoon (total area more than 300 ha) belong to Huong Phu and Hue City territory. There are two kinds of pond bottom, which are present in roughly equal proportions. Shallow bottoms are sandy or muddy sand while deep bottoms are muddy or sandy mud. In shallow bottom areas Halidube tridentata grow densely and in the deep bottom areas Chara zeylanica and Najas indica grow fast, especially when salinity is low. Depending on the season, there are about 25 species of seaweeds detrimental to Gracilaria verrucosa, mainly blue and green algae (Cyanophyta and Chlorophyta).

The lowest mean water temperature is in January (20.9°C) and the highest is in July (30.6°C). In the temperate months from February to April, the mean temperatures are 21.5°C, 24°C, and 26°C respectively.

Salinity varies with the seasons. In the rainy season from September/October to December/January, the salinity drops to below 5 ppt. In February, salinity begins to increase, and reaches its maximum of 32-35 ppt in July and August.

Mineral content (N and P) in the ponds varies depending on the time of fertilization and the level of development of the seaweed in the ponds. The highest concentration is usually in November and the lowest in April and August. In general, the Tam Giang brackishwater lagoon has average low mineral content. Surveys on other climatic factors in the region show that average annual rainfall is about 1900-3200 mm and distributed unequally, falling mainly in the period from September to November. The sun shines about 1700-1900 hours in a year. This is favourable for harvesting and drying seaweed, but seaweed grows slowly in the shallow ponds (under 30 cm deep) during the hot months.

Seaweed Culture Techniques

The main crop is grown in spring/summer (January to May), and a sub-crop is grown in summer/ autumn (July to September). Seedstock preservation is carried out during winter (October to December). The condition of the pond bottom is improved by submerged ploughing using a ploughing machine or a buffalo-drawn plough.

Seeding - using transplant seed from seedstock preserved during the winter or seed collected from natural habitats during the same period - is usually carried out in January/February. However, seeding may be carried out one month earlier or later depending on the arrival of warmer weather. Seeding is carried out a density of 200-500g/m^2 Additional manure is added only to the shallow bottom areas at about 2-5 tonnes/ha. Before planting, seedlings whether removed to a new place or gathered from nature are treated by pulse fertilization with inorganic fertilizers (N and P at a ratio N:P = 2:1) at 3-5 kg/tonne of seed. All water gates are closed in the rainy season to prevent a sudden drop in salinity. In the hot season, the gates are opened and closed daily (depending on the tide level) to allow more circulation of water in the ponds. During monthly periods of low tide, it is necessary to open inlet gates. During this period water should be drained only at night. Whenever foreign seaweeds develop and fully cover the G. verrucosa, harvesting should be combined with removing them so that G. verrucosa growth is not restricted.

Harvesting starts about 1.5-2 months after seeding and continues everywhere in the region up to the beginning of the rainy season. Time of harvest is also decided by the results of laboratory determinations of agar content and quality confirmed by field technicians. The amount of seaweed left after each harvest is 300-400 g/m^2. Seaweed is harvested by scoop net or by hand, then washed in the field and dried on dike embankments. It is transported to a freshwater area for washing and drying for a second time. Finally, it is pressed, packed and transported to the market.

Gracilaria sp. and Polycavernosa were cultured experimentally using four different methods: (a) field monoculture with insertion line technique (b) monoculture in pond (c) polyculture with seafish in sea cages and (d) polyculture with grouper fish in cages. The best results were obtained from field monoculture with insertion lines and seaweed monoculture in ponds. Insertion lines increased from 1 kg to 20 kg in one month, while pond monoculture grew from 10 kg to 60 kg in 3 months.

Before farming of agarophyte seaweeds can be undertaken on a commercial scale, it is necessary to establish the ecological parameters necessary for optimum production. This paper reports a study undertaken to determine these parameters for conditions prevailing in Thailand.. There are two genera of agarophytes with potential for commercial exploitation in Thailand. These are Gracilaria and the closely related Polycavernosa. Abbott (1987) has reported four species from each genus being present in Thailand. These are Gracilaria tenuistipitata, G. firma, G. irregularis, G. salicornia, Polycavernosa fisherii, P. changii, P. fasfigiata and P. percurrens.

Ecology of Natural Habitats

A study of the natural habitats of the agarophytes in Songkhla lake and Pattani lagoon was undertaken. The water temperature ranged from 20-35°C, with the optimum being 30°C. Salinity ranged from 10-35 ppt with an optimum of 25 ppt. The pH of the water ranged from 6-8, with the optimum being 7 (i.e. neutral). Agarophytes grow best in brackishwater. l-1.5 metres deep over a sandy/muddy bottom. Ideally conditions should be calm with little wind to generate waves which may lead to the seaweed aggregating on the leeward side of the lake.

Farming Trials

Experiments have been carried out to assess four potential methods of culturing agarophyte seaweeds.

At Ko Yo Island in Songkhla province, seaweed was monocultured using the insertion line technique. The seaweed cuttings were inserted into 50 metre lengths of 3.5 mm rope, and left to grow for one month.

At Da To village in Pattani province, the seaweed was monocultured in a 7 x 10 metre pond under 1 metre of water above a sandy/muddy bottom and left to grow for three months.

The other two experiments involved polyculture with fish in sea-cages. At Ko Yo Island in Songkhla province, the seaweed was placed in a 4 x 6 x 2 metre cage containing 300-350 sea fish and left to grow for one month. At Kantang in Trang province, the seaweed was placed in a 5 x 6 x 2 metre cage containing 300-350 grouper fish and left to grow for one month.

At the end of the growing period, the seaweed was harvested, cleaned with seawater and set out in the sun to dry (this usually took 2-3 days). Wet and dry yields were determined, and drying ratios calculated.

Results

Successful growth was achieved in all four experiments. The seaweed monocultured in insertion lines grew from an initial 1 kg to 20 kg after one month. The seaweed monocultured in a pond grew from an initial 10 kg to 60 kg after three months, i.e. an average growth rate of 20 kg/month. The seaweed polycultured with sea fish increased its weight from 10 kg to 20 kg in one month, and that polycultured with grouper fish increased its weight by 1620 kg in one month. The drying ratios calculated were 7: 1 for Gracilaria and 10:1 for Polycavernosa.

Quality and Price

To obtain the maximum price, dried seaweed should meet certain quality characteristics. It should be black in colour, be free of impurities such as shells, rocks and mud and have a moisture content of about 20 per cent. Dry seaweed meeting these standards should fetch a price of US $800 per tonne.

The effect of temperature, salinity, nitrogen, culture density and depth on the growth of Gracilaria tenuistipitata were investigated between April 1985 and March 1986 in outdoor ponds in Guangxi Province, South China. The mean annual growth rate was 2.4% per day. Under favourable temperatures of 20-30°C, daily growth rate may reach as high as 3.3%. Salinity had an obvious effect on growth and photosynthesis, and growth peaked at 21 ppt, with a broad plateau between 7-27 ppt. Growth experiments showed that a total nitrogen (NH,- N plus NO,- N) concentration of 4 m is sufficient to enable the plant to maintain a daily growth rate of 2.7%. The best growth of the plant was obtained at a culture density of 0.5-1 kg/m^2 and a culture depth of 30 cm in the pond.

Studies on Gracifaria are of special interest because of its increasing market value as a source of agar (ADB-FAO 1983), its importance in diet (Cordero 1984) and its significance as food in the cultivation of abalone (Chiang 1981). Recently, the demand for agar has steadily increased, while the supply of Gracifaria from natural populations has not. Consequently, cultivation of Gracilaria has begun to supplement harvests from natural populations (Mathieson 1975). The commercial beds of Gracilaria in China are so limited that the supply is unstable and cannot meet the everincreasing demands (Tseng 1980).

There are two varieties of Gracilaria tenuistipitata found along the coast of South China, G. tenuistipitata and G. tenuistipitata var. liui. Of these, the latter is mainly distributed in the north coast of Hainan Island and is far better than the former, because it grows faster and is adaptable to brackish seawater. G. tenuistipitata var. liui is, therefore, one of the most common species under pond cultivation in Guangxi and Hainan Provinces.

The cultivation area once reached 1500 hectares, producing some 2000 tonnes dry weight and played an important role in agarophyte cultivation in China. Unfortunately, the growth rate sometimes fluctuated greatly. In serious cases, most of the plants in the ponds died for no obvious reason within a few days.

The purpose of the work reported below was to determine the relationship between the main environmental factors and the growth rate and to find ways of overcoming the unusual deaths in order to stabilize and increase production.

Materials and Methods

The effect of temperature, salinity, culture, density, depth and total nitrogen content of the ambient seawater on the growth of G. tenuistipitata var. liui were studied concurrently in outdoor ponds (l-1.5 m deep) in Guangxi Province, South China and in the laboratory in Qingdao from April 1985 to March 1986. Vegetative fragments of G. tenuistipitata var. liui were collected from the outskirts of Haikou city, Hainan Island. Pond culture experiments were carried out at Baihutou station near Beihai City, Guangxi Province, South China as follows:

Temperature: Triplicate samples of Gracilaria fragments 5-10 cm long weighing 500 g were cultured in a rectangular net cage in a depth of 0.5-1 metre. Measurements of weight were taken every 15 days. growth rates were calculated and temperature monitored throughout the experimental period.

Salinity: Gracilaria were cultured in two ponds with different salinities using net cages similar to those used in the temperature experiments. The salinity of pond A was 30-34 ppt, and that of pond B was 24 ppt. Total nitrogen content of the two ponds was maintained at the same level by adding nitrogen fertilizer to pond A. Outdoor tank culture experiments and measurement of the effect of salinity on photosynthetic rate were conducted concurrently at Baihutou station and Qingdao Laboratory.

Nitrogen: Nitrogen nutrition experiments were conducted in three ponds in which the total nitrogen contents were different. Water samples were analyzed for ammonium-nitrogen (Gao. Zhang er al. 1980) and nitrate-nitrogen (Shi, Dai et al. 1980) periodically. Growth in fresh weight was also measured.

Culture, Depth and Density: Triplicates of 20 g of vegetative fragments were cultured in net cages at 30, 60 and 90 cam deptns respectively in a pond and increments in weight measured. At the same time. photosynthetic rates were measured by using the white-black bottle method. Culture density experiments were set up in two ecologically similar sites, each site covering an area of 30 m^2. The culture densities were 150 g/m^2 and 450 g/m^2 respectively. In each of the five plots selected at each site, a 0.04 m^2 net cage was used to culture Gracilaria at the same density. Fresh weight was measured every 15 days.

Prawn fever started in Indonesia in early 1980. South Sulawesi, with its large area of brackishwater ponds, was no exception. Seventy thousand hectares have been developed out of a potential 150,000.

When we started our company in 1986 and looked at supply and demand, we had our doubts about prawn farming. After all, if every one was going into prawn production, over supply and price declines were only a matter of time. Thus, we began to look at other resources and decided to go into Gracilaria farming.

Another consideration was that Gracilaria can be grown together with prawns, In fact, we found Gracilaria growing naturally in abandoned prawn ponds: further confirmation that the two could be cultured together.

Our company charter mandates a social function. Consequently, we wanted to involve low-income people of the province in seaweed farming.

Prawn prices have dropped in recent years, encouraging farmers to look to seaweed as an alternative source of income requiring little additional investment. Many shrimp farmers are heavily mortgaged and short of capital.

Gracilaria Development in South Sulawesi

As of 1986, there was no Gracilaria culture in south Sulawesi. Nevertheless, 1,400 tonnes of seaweed harvested from wild stocks were exported during the same year. This consisted mostly of Eucheuma spinosum. There were also orders for Gracilaria from Japan.

We decided to culture Gracilaria, but at the time we did not know much about culture technology. We found difficulty locating seed stock and began with material from Chile, originating from export quality parent stock. But finally, after surveying the coast of south Sulawesi, we located good material near Ujung Pandang at Tanjung Bunga.

This Gracilaria produced agar similar to that of the Chilean species. We also found similar seed stock in the Janeponto district. Eventually, we found that the species was Gracilaria verrucosa. Concurrently, in 1986, the University of Hasanuddin, at Ujung Pandang, the Regional Planning and Development Bureau of south Sulawesi and the Fisheries Department of south Sulawesi began investigating Gracilaria and we are now working together to develop pond culture of Gracilaria.

We set up our first pilot project in a 10 ha pond area at Tanjung Bunga. Our company provided seed stock, gave training seminars and paid for maintenance and operation costs. The pond owner only observed our activities at the beginning. Once we developed the technology, the owner took over and we purchased the seaweed from him. After harvesting Gracilaria in September 1986, we decided to transplant seed stock to Takalar. Unfortunately, the rainy season began in the area the following December and affected seaweed growth. But since it was the dry season along the coast of the Gulf of Bone to the west, we decided to take some of the Gracilaria from our initial site to Sinjai. Our activities at the original site were reduced to maintenance of the existing stocks and transplantation to new ponds. When we started seaweed farming, we were interested in locating seedstock, and were not really aware of the weather pattern in south Sulawesi. Now we know how to use this unique climatic pattern to maintain year-round production.

In addition to the challenge posed by climate, the company encountered problems in finding farmers willing to take up seaweed culture. In 1986 prawns fetched good prices, but some farmers unable to succeed with prawn culture were interested. However, they had doubts about marketing- who would buy their production? So we began work with these farmers, providing free seed stock and distributing brochures on farming methods. They still remained sceptical, but our persistance has paid off and farmers who seriously took up Gracilaria farming in their defunct shrimp ponds are gradually improving their living standards.

Export Development

Eight months after the company started distributing seed stock, we harvested 110 tonnes wet weight. Ten tonnes were set aside as seed stock and the remainder converted to 10 tonnes of dried seaweed.

After Japanese buyers accepted our product we sent off our first export shipment. Three months later, the company exported 10 tonnes of dry Gracilaria. In April 1988, we exported another 20 tonnes. Luckily for us, Gracilaria farming is picking up in several areas with some farmers developing new ponds for seaweed culture.

Now we have another problem-traders trying to muscle into our business and manipulate the price. But our good relationship with our farmers has protected us from such speculators, who are to be found in any business.

Our Stable Period

In the beginning we were dependent on farmers whose production output was variable. In order to stabilise production, we decided to develop our own ponds. This would enable the company to develop the best culture technology and operate to its own standards.

In December 1988, we rented a pond of 6.8 ha from a farmer. Although not too large an area, it enabled us to produce more seed stock for distribution to interested farmers.

By the beginning of 1989, our business started to show good prospects and we decided to buy a 20 ha pond in Takalar Regency. We can now harvest 20 tonnes of dried Gracilaria every two months which translates into 10 tonnes dry weight every month. Production has now stabilized.

A new factor entered the scene in March 1989, when prawn prices on the world market started falling. Now, farmers are getting more interested in Gracilaria because they are looking for alternative crops. Many farmers are asking us how to grow seaweed and are even willing to buy seed stock. Trade in seedlings of Gracilaria is building up, which reduces our costs since we have to supply less seed stock to get the farmers started. Increased availability of seed stock means production will increase.

Since the middle of 1989, we have been able to export 60-80 tonnes of Gracilaria per month. We are now developing a seeding strategy to enable year-round stable production.

Further Development

Because Gracilaria farming is taking off in south Sulawesi, the company will develop new sites in southeast Sulawesi and some sites further east. In southeast Sulawesi we are constructing ponds on a 200 ha site. Full development of the site will allow us to harvest SO tonnes of dry Gracilaria per month.

Culture methods will differ from those in south Sulawesi because we are working in new ponds. Acidic soil has forced us to wash out pyrites by cyclically pumping water out of the pond and refilling.

Based on our past success and bright prospects, the company will construct an agar processing plant in the region. It will consume 50-100 tonnes of raw material per month, hence our expansion into pond culture. Locating the plant in the production area will reduce transport costs and expand market outlets for the farmers.

Conclusions

Gracilaria farming in Sulawesi has a bright future. Involving local farmers reduces the company's costs, and while the company is developing new farm sites we will continue to involve them. But to stimulate further interest on the 'part of farmers, farm gate prices will have to be raised. We expect overseas buyers to raise their prices instead of having a situation similar to that of the falling price of prawns. We would also like to co-operate with international organizations to do more research. Raw material can be semi-refined in processing plants.

SMALL-SCALE AGAROPHYTE PROCESSING ASPECTS ON GRACILARIA

World statistics refer to fresh weight at harvest; the value for the fishermen. They should not be confused with export values of dried seaweed which may at the time have been about US $ 800/tonne. Of course industrial end products will be sold at hundreds or even thousands of times more than the raw material, but processing proceeds will not reach the fishermen.

Import and export figures are not all. We are happy to live in a time when life quality is also included in our demands, in this case health aspects. Around the Bay of Bengal there are still areas with protein deficiencies, keratomalacia (Vit.A), beriberi (thiamine), pellagra (niacin). ariboflavinoses (B2) and scurvy (C).

Gracilaria is famous for its high protein content. Some of the species are known for their quantities of, among others, provitamin A, thiamine and niacin. However, if seaweeds should go where they are needed, they should all be sent to areas afflicted with severe endemic goitre (the Himalayas and other highlands). (Michanek, 1979, 1981). Coastal populations never suffer from goitre. and vitamin deficiencies are few.

The health impact of seaweed as a food additive does not rest with its content of proteins, vitamins, minerals and trace elements. There is increasing literature on the capacity of phycocolloids to bind heavy metals and thus to decrease our body burdens of lead, cadmium, mercury. arsenic and even radioactive strontium. If ongoing research shows that we can rinse lead out of our bodies by regular intake of seaweeds, there will be an immense increase in the demand, in particular for the populations living and dying in the traffic exhausts of our cities. For this purpose agar and carrageenan are not so good as alginic acid. This may open a new use for resources of Sagassum, Turbinaria, Cystoseira and Cystophyllum.

In the extensive literature on Gracilaria, many papers deal with the question I started with: quality of agar. Let me cite a sampling of recent papers with aspects worth discussing by cultivators. From molecule chemistry we first learn that much sulphatc gives low gel strength., Second. we learn that a high curling intensity of the molecule chain gives a high gel strength and that this curling of rhe chains increases with increasing amount of 3,6- anhydrogalactose (Yaphe and Ducksworth, 1972).

For the cultivator this means that in tidal areas a Gracilaria growth which is exposed during long periods will produce galactans heavily charged with sulphates, as this radical plays a role in resistance against exposure. In subtidal specimens of the same transect. sulphates show partial absence and the molecules have a high portion of 3,6-anhydrogalactose (Bodard, et al., 1984).

Our third observation could be that 3,6-anhydrogalactose and consequently gel strength has a negative correlation to chlorophyll content - or in plain words. quality is not so good during sunshine periods (Liu, et al., 1981).

The genera Gelidium and Gracilaria are low in sulphate, while Chondrus, Gigartina and Porphyra are high (Bodard, Christiaen and Verdus, 1983). Within Gracilaria the species G. sjoestedtii has lower sulphate content and higher gel strength than G. tikvahiae, G. rextorii and G. verrucosa (Craigie et al., '1984). In a certain species different clones may have very great differences in agar composition and qualities. Strains with thin thalli have a more efficient uptake of nutrients than strains with thicker fronds (Lignell and Pedersen, 1989). Even different parts of the thallus show considerable variation (Craigie and Wen, 1984).

A fourth aspect is that, in general, gel strength is higher in agar from plants grown in nitrogenrich water and under good light conditions, while the yield of agar is higher from plants grown under poor conditions (Craigie et al., 1984, Lignell, 1988). This may be explained by the fact that plants in nitrogen-rich water grow faster than those in nutrient-poor water and therefore have a larger portion of young tissue. In young tissue the proportion of small cortical cells to large medullary cells is high (Craigie and Wen, 1984).

Chemical observation number five is that gel strength of agar deteriorates markedly with increasing content of 4-0-methyl-L- galactose, which simply means that aged tissue is inferior to young tissue (Craigie and Wen, 1984, Cote and Hanisak. 1986, Lahaye and Yaphe, 1988). From laboratory culture experiences we could note that Gracilaria is extremely sensitive to low concentrations of nitrogen (Edelstein et al. 1976).

Wang et al. (1984) observed that local plants doubled the fresh weight of their thalli in considerably less time than transplants.

What are the problems of seaweed culture? In one of the discussions at the Symposium of Useful Algae, which we know from the volume "Pacific Seaweed Aquaculture"(Abbot,Foster, and Eklund Eds. 1980), a participant stated that:

* Problem no. 1 is grazing and overculture.

* Problem no. 2 is weeds and epiphytes.

The problem of epiphytes, when there is no grazing, was studied by Brawley and Fei (1978). At the 13th International Seaweed Symposium in Vancouver last August there was a workshop on Engineering Aspects of Algal Cultivation, where various participants noted that:

* each time a culture system is scaled up, new problems arise

* production limits are probably set by the ability to agitate (circulate), rather than by self-shading.

* a given system can be made more economic by reducing the cost of production rather than by increasing productivity. (Applied Phycology, September 1989).

For the farmer there are many problems which we see nothing of in the literature, e.g.:

* Is an upgrading of living material through nitrogen starving feasible?

* For a sea farmer, is the way to gain more only to produce more or would it pay better to sit down and do a lot of hand-rinsing of epiphytes, other species and plastic fragments or to rinse a clay-rich harvest from a muddy bottom in clean water of the open sea?

* How much is lost in quality if there is a transport delay between harvest and processing?

* Does the buyer ask for quantities only, while the factory primarily asks for quality?

6

SMALL-SCALE AGAROPHYTE PROCESSING

INTRODUCTION

Gracilaria spp. are agarophytes or agar-yielding red seaweeds belonging to the class Rhodophyceae. They give rise to a hydrophilic colloid. agar, which on processing is insoluble in cold water but soluble in hot water. Other agar-producing seaweeds are Gelidiella, Gelidium, and Pterocladia. "Agar-agar" is the original Malay word for a food gel obtained from seaweed. Chemically, agar consists of neutral agarose and charged agaropectin. Agarose is a long-chain polymer of neutral galactose and its derivatives, while the charged agaropectin chain has some sulphated substitutes. A higher charged agaropectin chain leads to increased viscosity but reduced gel strength. Reducing sulphate content helps increase gel strength. According to Chandrkrachang and Chinadit (1988). the most important properties in determining the quality of agar are gel strength, gelling and melting temperature, sulphate and metoxyl content. clarity of the solution, and ash content. Apart from seaweed powder and dried seaweed from which agar is extracted, other product forms are strip-agar, agar flakes, and agar powder. High-grade agar is white, while light yellow is acceptable for lower grades.

HANDLING AND PROCESSING

Armisen and Galatas (1987) provided an excellent review of the handling and processing methodology for agarophytes. This paper while drawing on their work, attempts to summarise the technologies used. Preservation of seaweeds between the time of harvest and processing is very important to minimise spoilage, facilitate long distance transportation and extended periods of storage before processing. Before preservation, the seaweed is washed in sea or fresh water to remove adhering sand, mud, snails, barnacles or other foreign material.

Factors to be Considered in Agar Processing: The extract from the agarophyte has to contain the maximum possible quantity of agar from within the seaweed. In addition, the agar obtained should also have the best physico-chemical properties to satisfy the standards expected of the end-product. The normal sequence of steps in handling and processing agarophytes are dehydration, pressing, evaluation, removal of undesired products, pre-treatment prior to extraction and the control of molecular weight during extraction. There is a need to work with large volumes of dilute extracts. and to consider the economics of dehydrating.

Dehydration

Immediately after harvest, the seaweed is dried to a moisture content of less than 20%. Seaweed may be air-dried in the open, preferably off the ground, to keep it clean or dried artificially. Bamboo slats, plastic sheets on the ground, or concrete surfaces are used. Sun-drying also bleaches the seaweed. The seaweed must be sufficiently dry to prevent anaerobic fermentation resulting in spoilage or even carbonisation of the bales during storage.

The yield achieved for Gracilaria under experimental conditions in New Zealand is 40 tonnes of dried seaweed from 240-280 tonnes wet raw material. One tonne of the same dried Gracilaria yields 250-300 kg of agar (Hollings, 1985). In Taiwan, the drying ratio obtained is a similar 1:7 (Chen. 1978).

Preservation of Gracilaria by dehydration is difficult as enzymatic hydrolysis of the agar occurs even at relatively low moisture content. This rate is, however, variable depending on the species and its origin. Gracilaria harvested in Sri Lanka, India, Venezuela, Brazil and other warm waters has an agarose less resistant to enzymatic hydrolysis than the more stable Chilean Gracilaria. Agar in Gracilaria, which can undergo hydrolysis because of endogenous enzymes or the growth of Bacillus cereus, is less stable than that of Gelidium (Armisen & Galatas, 1987).

Pressing

The second step is pressing the weed in bales of about 60-100 kg with a hydraulic press, in order to reduce the volume and consequently transportation and/or storage costs. The seaweed may then be packed into sacks, either for export or sale locally. Alternatively, it may be ground into powder or subjected to agar extraction at the small-scale processor level.

Evaluation

Because of variations in harvesting and processing methods, chemical properties and product utilisation, it is not possible to consider seaweed and, therefore, the seaweed industry, as a homogeneous entity. Agarophytes from different growing areas should be evaluated for their agaryielding properties before processing, so that their potential can be assessed.

In principle, evaluation of agar properties involves combining some preliminary treatments with different extraction methods. An important consideration is sampling from a large growing area and the proper treatment of dirt-free samples until they are ready to be processed. A sample size of about 400-500 g of dried seaweed should be used. Provided good results have been obtained, a pilot laboratory which follows closely the actual operations that would take place on an industrial scale should be set up.

Next, aliquots of seaweed representing a homogeneous composition should be taken, and tested for moisture determination, pure seaweed determination, and extraction of agar. Moisture determination is caried out in a drying oven at 65°C.

With Gruciluriu, a sulphate alkaline hydrolysis pre-treatment is usually carried out to change the L-galactose

6-sulphate into 3, 6-anhydro-l-galactose. This is usually done by diffusion with sodium hydroxide solution (0.1M) for one hour at a temperature of 80-97°C taking care not to extract the agar. The agar extraction which follows is carried out while stirring the slurry at neutral pH without pressure for a period that varies depending on the type of Gracilaria. This can take several hours.

A manufacturer of good quality agar should be able to spot variations in yields in a laboratory trial before embarking on an industrial-scale production with a new batch of seaweed. The properties of seaweed from different areas vary greatly. This has resulted in the failure of many processing factories attempting to process batches of seaweed from different sources.

Removal of Undesired Products

In order to obtain the purest possible extract, seaweeds are generally hand selected and washed prior to alkaline treatment to eliminate a large quantity of foreign substances.

Pre-treatment

A variety of pre-treatments are available. Treatment differs depending on source, growth stage, environmental conditions and species.

Pre-treatment should achieve maximum desulphation, while avoiding yield loss through agar dissolving in the solution. The disadvantage is that long-chain polymers may break (as discussed in the following section on "molecular weight control") leading to a reduction of gel strength. An alternative technique to pre-treatment is strong alkali post-treatment at relatively low temperatures soon after agar extraction. Ry this means, relatively low-grade crude agar is refined to produce high-grade agar. Alkali post-treatment of traditionally extracted low-grade agar from Gracifaria has yielded a two-to-three-fold increase in gel strength, from 300-400 g/cm^2 to 700--1000 g/cm^2 (Chandrkrachang & Chinadit, 1988).

Shengyao et al., 1988, studied the effects of alkali treatment on agars from Chinese Gracifaria species using a cold concentrated alkali pre-treatment in which Gracilaria is treated with 32% NaOH at room temperature for five days. Agars extracted with sodium hexametaphosphate after cold concentrated alkali treatment were much better than those extracted with water or any other alkali treatment. The gelling and melting temperatures of agar treated by alkali increased while the viscosity decreased.

The 3, 6-anhydro-1-galactose content of agar isolated from alkali-treated Gracilaria was higher than that without alkali treatment, and the reverse relation was observed about the sulphate and galactose content. The increase in gel strength lay in the degree of the sulphate reduction and the 3, 6-anhydro-l-galactose increased.

In another study on pre-treatment by Minghe, 1986, Gracilaria was pre-treated with dilute alkali prior to extraction at a temperature of 78-80°C for three hours. The yield and gel strength of the agar produced this way were higher than that of agar produced with the usual concentration of sodium hydroxide solution.

Molecular Weight Control

Within the seaweed, agar is insoluble in both warm and cold water. Extraction has to take place within certain pH, temperature, and redox conditions. This results in some hydrolysis taking place, thereby increasing its solubility.

During this fractionation, it is necessary to minimise molecular weight reduction. Since all agar extraction works on the principle that agar dissolves in hot but not cold water, excessive molecular weight reduction would cause loss in yields for low-temperature water soluble agar. On the other hand, all molecules that have weights at the higher extreme will not be extracted and will remain in the seaweed.

Thus, a manufacturer has to develop a means of obtaining maximum yield of particle size in the mid-range. This translates mto a high gel strength with more uniformity in particle size and reduced losses at the extremes. It is difficult to modify the percentage of molecular weights dissolving below 20°C but the extraction of molecular weights dissolving above 125°C can be increased by raising the water temperature under pressure whenever the seaweed permits.

To further strengthen the agar product the acid concentration should be adjusted at pH 2-2.5, and for increasing productivity pH 4.0 - 5.0 was best. A pH of 3.5 is ideal, resulting in an agar yield of 21.6% and 650 g/cm^2 gel strength (Longchang, et al., 1986).

Dehydration of Dilute Extracts

Agar extracts of 0.8-I .5% are considered to be the optimal concentration for subsequent dehydration. The more the agar extracted, the larger the quantity of water required. Generally factories working with Gracifaria have a higher water consumption than those using other raw materials. Water consumption also increases when better quality agar is required.

Working with a 1% solution, 99 litres of water have to be eliminated to produce 1 kg of agar. Evaporation or precipitation is recommended where freezing/thawing is not practicable. Freezing and thawing the extract is employed both at industrial and cold country household level. Industrial freezing should be slow, to maximise both the growth of ice crystals and the separation of agar. This is usually followed by draining with a water-extracting centrifuge.

This freezing-thawing method. known as syneresis, followed by washing the gel matrix, also helps to purify the gel. Sulphated galactan portions of agar, salts, pigments and other organic compounds are soluble in the thaw water, and are partially removed from the insoluble agarose by filtration. Factories which use syneresis usually have high water consumption. Some Gracilaria agars form soft gels which on washing disintegrate into fine particles.

Similarly, low-temperature soluble agarose is lost when washing some agars. Thus, washing the gel matrix after the freeze-thaw cycle can only be performed with high gel strength agars or when alkaline pre-treatment is employed (Yaphe, 1984).

Small-Scale Agar Manufacture

Small-scale agar manufacturers with basic technology do not generally produce agar that meets international quality standards. There is little or no scientific control over processing, bacteriological contamination is usually too high, and prices offered too low. Consumption is usually local. Seaweed is washed and then boiled for agar extraction in water and, possibly, sulphuric acid for about four hours. Washing facilities differ from factory to factory.

To obtain a clear agar solution after extraction. some processors adopt pressure filters, or use high speed centrifugal machines; others depend on sedimentation. After filtration, the hot extract is poured into wooden boxes to gel. The gel is removed from the'boxes, sliced into strips, spread on mats and exposed to freezing cold weather to dry.

In order to produce agar strips, some use air-freezing rooms; others make use of brine tanks in small ice-making plants to freeze the agar gel strips inside ice moulds.

The processing of agar sheets is simpler than that of other products. it may be made in large or small plants. To produce agar powder from dried Gracilaria, the processing technology is similar until the freezing stage.

The frozen blocks of agar are then crushed, and the pieces washed to remove impurities. When the crushed pieces are melted, the agar is bleached and washed again before dehydration by low speed centrifuge. It is then dried to a moisture content of 20%. The pieces of dried agar are again crushed mechanically into powder and packed in polythene in quantities ranging from 100 g to 10 kg (Chen, 1978).

A simple cottage industry agar manufacturing process is common in the Philippines. Seaweeds (Gracilaria verrucosa, G. eucheumoides, Gelidium sp. and Eucheuma sp.) are washed with fresh water, sun-dried, and then re-soaked for 5-10 minutes. They are dried again until yellow, and then bleached in dilute vinegar until the colour turns olive green. They are then further dried until they become light brown.

Extraction is done by boiling in vinegar or even sulphuric acid with constant stirring. It is strained in cheese-cloth while still hot and cooled to gel at room temperature. Once set, the agar is cut into bars or strips, sprinkled with salt and ice, and frozen. It is then thawed and dried at room temperature or under the sun, after which it is ready for the local market (Guzman & Guiang, 1987).

Comment

It appears that the seaweed industry in the Bay of Bengal region suffers from a lack of marketing information and a need to improve technical know-how on processing. The establishment of processing facilities within the region would help stabilise the industry. The large exports of dried agar-yielding seaweeds, and the exports of high-valued agar products from importing countries such as Japan, should further encourage agar processing. Home industries should be developed with an eye towards product development, and also to suit the local demand.

PRODUCTION OF AGAR FROM SEAWEED WITH SPECIAL REFERENCE TO INDIA

World Production

Marine algae provide a rich and diverse source of raw material for the production of seaweed gums, polysaccharides that find wide application in the food, pharmaceutical and industrial sectors. The three most important polysaccharides, in terms of volume and value, are sodium alginate (and its derivatives), carrageenan and agar. A recent estimate (Anon. 1988) has put annual world production of agar at between 7,000 and 10,000 tonnes. The production of carrageenan and alginates, by comparison, is approximately two and three times as great, respectively. Seaweed gums: Annual world production (tonnes)

Agar 7,000-10,000

Carrageenan 12,000-15,000

Alginates 22,000-25,000

According to Armisen and Galatas (1987), of around 7,000 tonnes of agar produced in 1984, approximately half came from Gracilaria, the remainder coming mainly from Gelidium. The breakdown by country was as follows:

Agar: World production, 1984 (tonnes)

Japan 2,440 Taiwan 275

Spain 890 Argentina 200

Chile 820 Indonesia 150

S. Korea 600 China 140

Morocco 550 Others 300

Portugal 320 TOTAL 6,685

Indian Production

Observations in India and discussions with all but a few producers, indicate that current agar production is about 75 tonnes annually. If the expansion of output planned by several companies is realised, and the commissioning of other factories now under construction also comes about, total production could almost double within the next few years to around 140 tonnes per annum. Collection of seaweed destined for agar production, as also that of Sargassum and Turbinaria for production of alginates, is confined at present to the southern part of the Tamil Nadu coastline, between Cape Comorin in the south and the peninsula that stretches out towards Sri Lanka and forms the Gulf of Mannar.

Indian Agar: Species of Seaweed Used

Gracilaria

G. edulis (1).

G. verrucosa (2)

Gelidium Gelidiella

Gelidium spp. G. acerosa

G. crassa

G. corticata

G. multipartita (3)

Notes: (1) Syn. G. lichenoides

(2) Syn. G. confervoides

(3) Syn. G. foliifera

Of the Gracilaria species, G. edulis, collected from the waters off the mainland coast and those surrounding the off-shore islands, is the principal agarophyte. G. verrucosa, found in less salty estuarine areas, is used by a few agar producers. The other Gracilaria species are collected more by accident than design and do not form any significant part of the raw material utilised. After landing the seaweed, the collector sells his haul to an agent who then dries it and sells it to the processor. A small agent may subsequently sell it to a larger one. Seaweed prices paid both by the agent to the collector and by the processor to the agent reflect the higher quality of the agar obtained from Gelidium and Gelidiella compared to that from Gracilaria.

Indian Agar: Seaweed Prices

Seaweed

Gelidium/Gelidiella

Gracilaria

Notes: (1) Wet weight

Price paid (Rs./tonne)

Agent to collector(') Processor to agent(2)

1800 5,000-8,000

600 2,500-3,500

(2) Dry weight

Unlike Gracilaria from other sources, Indian material appears not to respond to alkali treatment as a means of increasing gel strength. Agar derived from Indian Gracilaria typically has a gel strength in the range 100-150 g/cm^2, while that from Gelidium or Gelidiella is around 300 g/cm^2. Two types or grades of agar are manufactured in India: food grade, which is usually produced in mat form, and IP grade, which conforms to Indian pharmacopoeia standards and is usually sold in powdered form.

For food use, paleness of colour is invariably considered more important than gel strength, and Gracilariu alone (which is cheaper and easier to bleach than Gelidium or Gelidiella) or a mixture of Grucilaria with Gelidium is commonly employed as the raw material. In a few cases, or where high gel strength is required, Gelidium alone is used to produce food grade agar, often in the form of shreds or so-called 'individuals' (strands of larger dimensions than shreds). IP grade agar is produced wholly or mainly from Gelidium or Gelidiella. Prices of the agar also, of course, reflect the raw material used and, for IP grade, the more stringent processing requirements. Within the Muslim community, demand for agar, and therefore also its price, is high during the Ramzan season.

Indian Agar: Prices for Agar Obtained by Producer

Grade Type Price (Rs./kg)

Food Mat 140-160

180-210 (1)

200-300 (2)

Shreds/'Individuals' 270-280

IP Powder 300-400 (3)

Notes: (1) During Ramzan season

(2) Prices quoted by a few producers

(3) One producer quoted Rs.600/kg

The factories that produce the agar are located for the most part in Tamil Nadu, in close proximity to their raw material source. A few, however, are more distant from the seaweed belt, in Kerala and Andhra Pradesh. The term 'factory' is used in the broadest sense to cover everything from the smallest family unit producing 2 kg of agar per day to the larger units producing 60 kg. Recently constructed factories plan to produce up to 100 kg/day.

Indian Agar: Production Levels

Almost without exception in India, the same basic method of processing is followed by all producers: acid treatment of the cleaned seaweed followed by hot water extraction; primary dewatering and purification of the agar by means of a freeze-thaw cycle; bleaching; and sun drying. Some adjustment is made according to whether Gracilaria or Gelidiella is being processed or whether it is G. edulis or G. verrucosa.

Initial handling and cleaning of the seaweed involves laying it out in the sun to dry and to bleach, removing epiphytes and other foreign matter by hand, and washing or soaking the weed several times in water. The seaweed may or may not be dried in between washes.

Acid treatment, to soften the weed in preparation for extraction, is accomplished by immersing it in cement tanks containing dilute hydrochloric acid for periods from 10 to 30 minutes, depending upon the species of seaweed.

After washing the seaweed free of acid, it is boiled in water at normal pressure without the addition of chemicals. Small units may employ direct heat by wood fire to boil the water; otherwise it is done by the use of steam. The extraction vessel may be fabricated of wood, aluminium or stainless steel. The length of time needed to extract 'the agar is dependent on the quality and nature of the raw material, but is usually somewhere between 1.5 and 3 hours. Gelidiella requires a somewhat longer time than Gracilaria. Yields of agar are around 10%.

After filtering, the extract is run into aluminium trays and allowed to gel. The trays are then transferred to a freezer, where they are kept, usually for 20 to 24 hours but sometimes longer. After removal from the freezer and thawing/draining, the crude agar gel is washed and then bleached by immersing briefly in hypochlorite solution. After washing again, the gel is laid out on mesh screens in the sun to dry. For IP grade agar, particular care is taken during handling and drying to avoid contamination by specks of dirt and other foreign matter. The sun-dried agar may be further dried in a hot-air drier.

PROSPECTS OF AGAR INDUSTRY IN INDIA

Utilisation of seaweeds in India for the extraction of soda ash, alginate and iodine started during the II World War. Although the importance of seaweeds was realised during this period, the production of agar did not start until the 1960's. The export of seaweeds continued until 1975 when, in order to meet the requirements of the local agar industry, the Government of India banned the export of seaweeds. However. it should be noted that the local industry does not produce sufficient agar to satisfy the ever increasing domestic demand. Consequently, India imports a large quantity of agar. The important agarophytes of India are Gelidiella acerosa, Gracilaria edulis, Gracilariacr assa, Gracilaria corticata and Gracilaria folifera. The seaweeds used for commercial extraction of agar are Gelidiella acerosa, Gracilaria edulis and G. crassa..

Gracilaria edulis is widely exploited for industrial utilisation in India. Collection is possible throughout the year around the islands in the Gulf of Mannar, Tamil Nadu. The, Gracilaria collected invariably contains various other plants. Fresh seaweed is sold at the rate of Rs.0.50/kg. When dried, the yield is 15% of the fresh material Commercial dried material is 60% pure, with a moisture content of 22%.

Since 1983, Gracilaria crassa has been collected from Pampan, Vedalai and Kilakarai, Tamil Nadu. G. crassa grows in shallow areas attached to pebbles and stones; collection is done by hand picking. Only negligible quantities are harvested, and then only when there is no collection of G. edulis. It fetches a price of Rs. 1,000 per tonne (dry weight). Agar is a complex mixture of polysaccharides obtained from certain species of red algae. It is mainly a mixture of two polysaccharides, agarose and agaropectin. Humm (1951) and Yaphe (1959) have defined agar as a gel-forming substance soluble in hot water and requiring a 1% solution to set as a gel on cooling.

The yield of agar and its gel strength vary from species to species and also on the method of extraction. Processing conditions play a significant role in the quality of the end product. In India, Bose et al (1943) soaked dried seaweeds in water for 18 hours prior to treatment with acetic acid. Chakiaborty (1945) did the same, but used the process of freezing and thawing for purifying the agar gel, and used activated carbon to decolourise the gel. Karunakar et al (1948).

and Joseph and Mahadevan (1948) purified the gel by soaking it in water of low salt content for 96 hours and finally washing under pressure. After freezing, the gel was dried with acetone. Thivy (19.51) soaked the seaweeds for 24 hours prior to extraction. Kappanna & Rao (1963) studied the method of Thivy, and found that soaking and extraction under pressure reduced the quality of agar gel. Srinivasan and Santhanaraj (1965) washed the seaweeds in sea water and then in fresh water several times, and dried them in the sun until they were completely bleached. Extraction was carried out by boiling the seaweeds at pH 6 and the extract was finally filtered and frozen. A cottage industry method for agar extraction from Gracilaria edulis was worked out by Thivy (1958). Umamaheswara Rao (1970) has given a comprehensive account of aspects of Indian seaweeds and their utilisation. Much work has been reported on the chemistry of Indian seaweeds (Pillai, 1955a. 1955b). A comparative study was made by Chennubhotla et al (1977) on the yield and physical properties of agar from different agarophytes.

AGAR PRODUCTION ADAPTED TO RURAL AREAS

Agar-bearing seaweeds, agarophytes, are naturally abundant along more than 2,600 kilometres of Thailand's coastline. Production of natural stock is rather seasonal along the Gulf of Thailand and less so along the Andaman Sea coast. The cage culture of marine fish has become a wellestablished industry, and agarophytes are appearing as fouling organisms attached to the net cages. These potentially valuable resources remain unexploited in most areas. However, Thailand exports raw seaweeds to developed countries and imports processed agar in increasing quantities every year. As a result, Thailand faces a trade deficit in agar. To reduce the country's dependence on agar imports, the Royal Thai Government recently started a vigorous drive to produce and process seaweed.

It focussed on stimulating private investment in seaweed processing for the local manufacture and marketing of agar, while at the same time creating employment for the impoverished coastal population. Within the framework of the programme, the Biopolymer Research Unit (BRU) was established and equipped to work on the identification of seaweeds and the analysis and process development of phycocolloids in Thailand. As a result of long-term fundamental research, the BRU has developed a simplified method of agar extraction from many types of Gracilaria and Polycavernosa species in Thailand. This simple technique is suitable for introducing small-scale crude agar production to rural areas. These can be used locally in food and tissue culture. The crude agar produced in rural areas can be collected and further refined with more sophisticated techniques in a central process&g plant to produce high-grade agar for different applications. This appropriate technology can be transferred to the rural people to make them aware of the wealthy seaweed resource on their doorstep.

MATERIALS AND METHODS

Agar-bearing seaweeds were collected from different locations in Thailand and identified by Dr I Abbot of the Department of Botany, University of Hawaii. Some former members of the genus Gracilaria have been moved to the genus Polycavernosa. The natural stock of seaweeds collected from Songkhla Bay were identified as Polycavernosa fishery.T he agarophytes growing attached to the net of marine fish cages along the canal of Trung were Polycavernosa changii while the samples collected from pond culture in Pattani were identified as Gracilaria tenuistipitata.

The collected seaweeds were placed on a clean surface and sun-dried for 2-3 days. Dry agar-bearing seaweeds are rather dark in colour and can be kept for a long time before extraction.

The procedure starts with cleaning the seaweed by washing with fresh water 2-3 times and then soaking for 2-3 hours. The cleaner the seaweeds, the cleaner will be the agar produced. After cleaning it is dried in the sun; the colour becomes paler, usually yellowish or light-brown. Before extraction, the dry and clean seaweeds are soaked in water in a ratio of 1:20 by weight for 2-3 hours until full hydration is reached. When the seaweed is completely swollen, the mixture is gently boiled in a water bath for 45 minutes. The seaweed is then ground using a traditional stone mortar or an electrical blender. The slurry is boiled gently for another half an hour. It is then poured into a simple hand-operated press which is lined with a two layered cloth bag, and which has been preheated by rinsing with hot water.

This simple press is a traditional local design used for extracting oil from coconut. After pouring the seaweed slurry into the press, the cloth bag is firmly tied and pressure is quickly applied. The filtrate flows into the tray, and the residue in the bag can be used as fertiliser. At this stage, desserts can be made by adding syrup and flavours. The filtrate is left to cool to allow the agar gel to set. The agar gel is put into a freezer for two days. The tray is then thawed at room temperature and the liquid is removed. The agar residue is sun-dried and stored for further use. If no freezer is available, the agar gel is put into a thick cloth bag and gently squeezed, either in the press or by putting a heavy stone on top of the bag overnight. The damp agar residue is then sun-dried and stored.

A mobile team of BRU staff arranged the transfer of this simple technology to the rural villages of Songkhla, Satul and Trung provinces during April 1989. Ordinary kitchen tools--cooking pot, blender, gas stove and coconut press there-were put into a compact package which could be adapted for mobile training facilities in three different rural areas during a 5-day operation. More than 100 local villagers participated in the seaweed production and processing training programme. Extension aids comprising video tapes, slides, leaflets and booklets were distributed.

RESULTS AND RECOMMENDATIONS

The crude agar extracted by this simple method is rather soft, but it can be used for local food. If a stronger gel is needed, more agar can be added. The mobile training programme appeared to be a great success and beneficial to the local population who responded well. Requests for more training were received from leaders of local communities. The crude agar produced in the rural areas can be used locally, or else collected for further refining to produce high-grade agar.

The planned production of crude agar in the rural areas will allow some of the benefits of seaweed processing to be channelled to the usually impoverished coastal population. By raising their interest in seaweed collection, culture and processing, the raw material requirements of a central commercial agar reprocessing plant will be established.

7

Extraction of Agar from Gracilaria Edulis as a Village Level Technology

A simple method of agar extraction, which has been developed in Thailand, may be suitable for adaptation as a village level technology in fishing villages in Tamil Nadu, India. Preliminary results from trials using the'modified method indicated an agar yield of 12.0 - 24.5% depending on the heating time and the method of heating. Suggestions for further research are discussed.

INTRODUCTION

Seaweed collection is an important source of income in fishing villages along the coast of Ramanathapuram district in Tamil Nadu, India. Agarophytes such as Gracilaria spp. and Gelidiella spp. are collected through most of the year by men and women. The seaweed is sold ashore for a few rupees per kilogramme to local agents who, in turn, dry the seaweed before selling it to processing factories for agar extraction. Agar is sold within India and is used mainly in the food industry.

The BOBP's Post-Harvest Fisheries Project is looking into the possibility of agar production being adapted as a village level technology.

By selling agar instead of seaweed, the villagers may be able to increase their income. A simple method of agar extraction has been developed by the Biopolymer Research Unit of the Srinakarinvirot University in Bangkok, Thailand (Chandrkrachang & Chinadit, 1988). This method produces a crude agar which could be upgraded and used in the domestic food manufacturing industry in India. With a few adaptations, this small-scale extraction method might be suitable for use in the villages of Tamil Nadu.

Since June 1989, trials to produce agar from Gracilaria edulis have been conducted in a field laboratory. The main purpose has been to determine the maximum agar yield using this extraction method, and to estimate the amount of seaweed which can be processed per day. This paper describes the agar extraction technology and the modifications in the technology required to adapt it for use in Indian villages. A summary of preliminary results and suggestions for further research are presented.

MATERIALS AND METHODS

Materials: To conduct agar extraction trials. the following materials were used: clean seaweed; fresh water; tubs to soak and wash the seaweed; a pan (50 1) to boil the seaweed; wooden spoons; a kerosene or wood-fuelled stove; a screw press; two planks and some heavy stones; filter cloths; trays and a platform to dry the agar in the sun.

All materials were purchased locally. The screw press was a larger version of the coconut press used in Thailand.

Seaweed collected from natural seaweed grounds, or from a seaweed farm, was cleaned and fully dried in the sun, so that it could be stored for some time. Before processing, the seaweed was washed and then soaked in fresh water for several hours until it felt soft. It was then heated in fresh water, the time depending on the amount and variety of seaweed. After heating, the seaweed was filtered through a cloth with the help of the screw press. The filtration had to be done quickly, and the screw press pre-heated with hot water to prevent the agar solution from setting during the process. The agar formed a gel after cooling. To remove the water from the agar, the gel was enclosed in a thick filter cloth and put under pressure either in the screw press or between two planks weighted with heavy stones for larger quantities. This process takes at least half a day, after which the agar needs drying in the sun for several days.

EXTRACTION TRIALS

Trials were conducted with small (100 g/l) and large (lkg/ 101 and 2kg/20 I) samples of seaweed. The small (100 g) samples of seaweed (ground or un-ground, bleached or unbleached) were heated in one litre of water for 1, 1 .5, 2, or 2.5 hours in a water bath, or heated directly. The larger samples were heated directly for 2, 3 or 4 hours. All trials were conducted with samples from the same batch of seaweed, which was collected from a seaweed farm and fully dried. Water was added if substantial evaporation occurred during heating, and the water temperature was kept at 90 to 95°C. The seaweed residue was heated for a second time in 500 ml water for 20 minutes (100 g samples), in 4.75 1 for 1 hour (lkg samples) or in 10 1 of water for 1.5 hours (2kg samples). In addition to these experiments, samples from the same batch of seaweed were sent for evaluation of agar yield and gel strength to the Biopolymer Research Unit in Thailand, the Central Salt Marine Chemical Research Institute in India and the Central Marine Fisheries Research Institute in India.

An economic model has been established for seaweed farming and agar extraction at the village level. The experience gained by the BOBP pilot farms near Mandapam, Tamil Nadu, India, provides the quantitative data. At a production level of 800 kg dry weight of Gracilaria edulis per plot, it is concluded that an agar-selling price of Rs.125 per kg is required, assuming a 1.5% yield of agar, for the project to he profitable. If seaweed production falls to 500 kg. an agar price of Rs.150 per kg is needed, together with an agar yield of 16%. The model shows that the project may still be profitable after fencing each plot to prevent grazing by rabbit fish.

During the last two years BOBP has operated small-scale pilot farms for Gracilaria edulis near Mandapam on the coast of Tamil Nadu, India. Recently, small-scale extraction of agar has been evaluated. A model for assessing the economic viability of the activities has been established. This paper describes and assesses this model, and then illustrates some of the economic relationships between key technical and economic parameters. It should be stressed that the model is an example, and that there are several alternatives that could be equally or more appropriate.

Structure

The economic model is intended to be dynamic. permitting simulation of different scenarios and their impact on the overall profitability of the project. The main profitability criterion utilised is internal rate of return, but to avoid academic discussions on the pros and cons of different criteria, net present value and pay back period have also been included. The model focuses on actual annual cash flows, rather than any accounting-related measurements of returns. To simulate different scenarios of possible outcomes of the project, a point of departure, here called "base-case", has been established. This includes the most likely values of all variables involved at the time the model was created.

Assumptions

The model is intended for a 0.1 hectare plot of G, edulis culture. All equipment for culture and drying of the seaweed, as well as extraction of agar, is assumed to be used exclusively for this plot. Because of grazing problems encountered in Mandapam, a major investment in fencing the plot has been included. The option of using natural seaweed stocks for agar production has not been considered, since the natural resources in the area have been severely depleted by harmful harvesting methods. Natural resources are, however, still harvested and sold as wet seaweed. Cultured seaweed, sold directly as seaweed, is obviously not cost competitive with that from natural resources, so this option has not been examined.

The opportunity cost of labour has been fixed at Rs.20 per man day, the salary presently paid to fisherfolk participating in the pilot culture. The amount of labour required for planting and maintaining the seaweed farms has been considered as fixed, i.e. it does not vary with the amount of seaweed produced.

The labour utilised to harvest and dry the seaweeds is expected to vary with the amount of seaweed produced. For extracting the agar, it is estimated that one man day will be needed for every 4 kg of dry seaweed processed.

The economic lifespan of individual items has been estimated to equal their expected technical lifespan. All replacement investments and future cash flows are expressed in fixed year one rupees The income statement contains a specification of sales revenue, all operating costs, and the annual depreciation of investments.

The final table of the model, containing the components of net cash flow per year, is the basis for measuring the profitability of the project. Using net present value, the annual net cash flows have been discounted at an annual discount factor of 15% and 25%. The internal rate of return expresses the discount factor at which the net present value of the project is equal to zero. Finally, the payback period. though crude, informs the prospective investor how much time it will take to recover his initial investment.

A Critical Assessment

Among the assumptions with a high level of uncertainty is the projected seaweed production figure. Pilot farms in Mandapam have not been able to produce anywhere near 800 kg dry weight per plot per year. This seems to be due mainly to grazing and to wind conditions in the open sea culture. If these problems can be overcome, however, it is still felt that an annual production of 800 kg dry weight of G. edulis, representing around 2.7 kg wet weight per metre of line per year, is technically feasible.

The technical parameters for agar extraction have been established during four months of practical trials. and consequently should be considered to be fairly reliable. The agar selling price is definitely on the high side. At present, local food grade agar is sold for between Rs.150 and Rs.250 per kg, the price varying because of festive seasons. It can be argued that the opportunity cost of labour should be adjusted downwards during the lean fishing and seaweed collecting periods. The quantity of labour needed is based on two years of experience, and should therefore be fairly accurate.

The expected lifespans of the various investments are naturally hard to predict, given the long planning horizons involved. Residual values of the assets at the end of the period have not been included, since they can be expected to be very small for the items not fully depreciated. Interest costs and interest earnings have been omitted from the model. In "base-case", the initial financing required could be repaid within three years, and interest earnings could be generated on the positive cash flows of subsequent years. It could, however, be argued that the positive cash flows from year four onwards would most likely be used to increase standards of living, in which case little or no interest earnings would be generated. There is also a high variability in the cost of financing, depending on the credit source. A subsidised bank loan at 4-5%, per annum would naturally result in a significantly more profitable project than a long term credit from the informal sector at considerably higher interest rates.

In summary, the optimistic predictions on primary seaweed production and agar price are only partly offset by over estimating the opportunity cost of labour, not including residual values, and a probably unnecessarily expensive fencing method.

Some Economic Relationships

The intention of the model has been to assist in quantifying the project in economic terms. It was felt that there was a need to assess the impact of alterations in cost, revenue, and output levels on profitability. The parameters identified as crucial for economic viability were seaweed production level, agar yield, agar selling price and fencing costs.

In the figures presented below, internal rate of return has been used throughout to measure profitability. The project should be considered economically viable when the internal rate of return exceeds the opportunity cost of capital. The opportunity cost of capital equals the return forgone by investing in the project rather than the best alternative project of equivalent risk.

GRACILARIA CORTZCATA : A POTENTIAL SOURCE FOR BIOACTIVE SUBSTANCES

During the last few years, interest has been awakened in marine biodynamic substances. Antibacterial potential has been detected in a number of macroscopic marine algae from all over the world. Distribution of these compounds in Rhodophyceae has been reported by Gerber et al. (1958), Glombitza (1970), Glombitza and Heyser (1971), Glombitza and Stoffelen (1972), Glombitza et al. (1973), Glombitza et al. (1974), and Hornsey and Hide (1974). Agar extracted from Gelidium and Gracilaria exhibited a marked inhibitory effect on the growth of the influenza B and mumps viruses (Gerber et al. 1958 and Takemoto and Spicer, 1965). This has been attributed to the agar galactan units (Nigrelli et al. 1967). In the late 1970s a study was undertaken on the antibacterial activity of Indian marine algae. This paper describes a part of that study, in which Gracilariu corticutu was identified as a potential source of bioactive substances.

Materials and Methods

G. corticuta, collected from Diu on the coast of Gujarat, was thoroughly washed to remove extraneous matter and epiphytes and then frozen until required. A method has been standardised for extracting plants using different solvents (organic and aqueous) and temperatures (20°C and 50°C). Staphylococcus aureus, Bacillus meguterium (gram +), Escherichia coli, Proteus vulgaris, Shigella sonnei and Salmonella typhosa para A (gram -) were used as test bacteria. Bioassay was by the agar plate diffusion test. The degree of sensitivity of the test organisms was determined by measuring the zone of growth inhibition. Four different fractions i.e. fraction A, fraction B, phenols and pigments were prepared from G. corticutu using the method reported by Parekh (1978). An in vitro comparative study of these fractions was conducted with known antibiotics (supplied by Bharat Laboratories, Bombay) against all the test organisms.

The antitubercular activity of different concentrations of fractions A and B was evaluated at Shri K. J. Mehta T.B. Hospital, Amargadj, Madhya Pradesh.

Results and Discussion

The antibacterial activity of crude extracts of G. corticutu against gram + bacteria is shown in Table I. The diethyl ether, acetone and ethanol extracts exhibited better activity than the chloroform extract. Hot and cold aqueous extracts and ammonium sulphate extract were not effective against any of the test organisms. The heat stability of the active principle was demonstrated, there being no difference in activity by acetone extracts prepared at 20°C and 50°C. * Paper presented at the seminar by Dr. V.D. Chauhan, CSMCRI.

MARKETING AND MARKET POTENTIAL FOR AGAROPHYTES AND EXTRACTS IN BOBP REGION

Out of nearly 260 species of seaweed growing in the coastal waters of Sri Lanka, only two species are exploited commercially at present: Gracilaria verrucosaa nd Gracilaria edulis. Various factorslike non-availability of resource data, deficiencies and difficulties in collection and marketing, and lack of adequate investment or interest from the industrial sector-hamper development of a seaweed production and processing industry. The export of dried Gracifaria has been highly variable partly because of the unsettled conditions in the country.

General

Sri Lanka has a coastline of approximately 1700 km along which many varieties of marine algae are found. Most of these marine weeds are of limited economic potential, except for a few species of Chlorophyceae, Rhodophyceae and Phaeophyceae. However, only two species of algae are commercially exploited in Sri Lanka at present, namely Gracilaria verrucosa and Gracilaria edulis. Since the mid-19th century, various studies have been carried out on taxonomical aspects and on the commercial importance of the seaweeds of Sri Lanka (Sigmond, 1841; Harvey, 1853; Svedelius, 1903: Tressler, 1923; Deraniyagala, 1933; Boergesen, 1936).

Present Status of the Seaweed Industry

In the 1970s about 50-100 tonnes of dried Gracilaria were exported annually from Sri Lanka. The quantity exported in 1972 was around 50 tonnes with an export value of Rs.76,000 (FOB). In 1986, exports increased to 150 tonnes, of which 70 tonnes were from the Kalpitiya area and the balance from Trincomalee. Over the last two years (1987 and 1988) there was no harvesting of Gracilaria in Trincomalee due to the unsettled conditions which prevailed in the area during this period. There are no figures available with the Sri Lanka customs or the Export Development Board for 1987/1988. However, according to one of the main exporters there was a sharp drop in exports in 1987, with only 10 tonnes of dried Gracilaria being exported. In 1988 this was further reduced to 5 tonnes. However, it appears that there is a renewed demand by the importer countries, as evidenced by an export order from Japan for 20 tonnes of dried Gracilaria received by a Sri Lankan exporter recently.

During May to November, wild Gracilaria grows well in the Puttalam lagoon. Harvesting is normally carried out during the phase of vegetative growth from May to November. This coincides with the south west monsoon period and also the season for prawn fishing in the lagoon. This coincidence could possibly be a reason for the lack of interest shown by fishermen for harvesting seaweeds during this season.

The shrimp fishery is the main livelihood of people in the Puttalam lagoon area. They fish in the lagoon with drag nets and cast nets during the monsoon season. Their average daily income derived from the shrimp fishery is around 200 to 250 rupees (6-8 US$). A boatload of fresh seaweed weighing'around 200 kg, with a dry weight of around 30 kg can be collected by two people in 5-8 hours. The collectors receive 350 rupees (10-11 US$) for 50 kg of dried seaweed from agents who represent export companies based in Colombo. As a maximum of two operations only are possible in a day, the total income rarely exceeds 700 This income has to be shared with all those who were involved in the collection. A recent NARA socio-economic survey (1988) revealed that only one quarter of those engaged in seaweed harvesting have their own boats. Furthermore, the collectors have to hire boats from outside sources and also have to pay for fuel. Thus, they end up with a very small profit margin from these operations. In comparison, shrimp fishing is more attractive as it gives a better and more reliable income, due to the steady market potential of shrimp.

Sargassum is not exploited on a commercial basis at present. However, there have been several enquiries about the purchase of Sargassum from importing countries such as Japan. Because of uncertainties in the supply of raw material, exporters are not in a position to undertake orders. It appears also that difficulties in detailed identification of the plant have posed difficulties to these exporters.

Marketing Network

Seaweed collected by the boats is unloaded on the beach and allowed to sun-dry for 4-5 days. The dried weed is tightly packed in gunny bags and transported by cart to the purchasing centres. Even though some cleaning is done by the collectors during sun drying, further cleaning is carried out at the purchasing centres. Here, the seaweed is sieved to remove particulate matter such as sand and sea shells, and then stacked to await transportation to Colombo. The middleman gets around Rs.10-12 for the dried seaweed. Further cleaning is carried out by the exporters prior to bailing for shipment. 'A bail of dried seaweed weighs approximately 150 kg.

A very small percentage of the dried seaweed is sold locally. Retail packs, weighing 50 to 100 g, are sold at most supermarkets, pharmacies and grocers in Colombo and its suburbs. Packeting of seaweed is carried out by traders who get their supplies from their agents. These agents visit the producer areas periodically and purchase the dried weed direct from the collectors. A retail pack of 100 g is sold at Rs.20-30, with the retailer getting a 15-20% commission. According to most Colombo traders, packeted seaweed is a slow-moving item except during the Islamic festive season, when demand is good. However, according to these traders, over the last few years there has been a growth in seaweed sales and it is used increasingly in local cuisine. They also believe that the local seaweed is superior to the imported, as evidenced by consumer preference for the former. Seaweed is a popular item of food in producer areas as well.

A recent socio-economic survey carried out by the National Aquatic Resources Agency (1988) revealed that 9 out of 10 people living in the Puttalam lagoon area use Gracilaria for domestic consumption. The most popular preparation is a porridge made by washing the dried seaweed several times and cooking it in water for 15-20 minutes. The thick soup obtained is sieved using a cloth strainer or wire mesh sieve, and coconut cream and lime are added to taste. The villagers believe this porridge to be highly nutritious and it is considered a must during the fasting season. The porridge is sold at kiosks in the Kalpitiya area at Rs.2/glass, and enjoys a very good demand among the local fishermen who rarely miss taking it on their return from the sea around noon time. According to a Kalpitiya boutique owner, the "commercial" recipe for the porridge includes 2 kg seaweed, three limes and 600 g sugar giving 30 glasses of the porridge. Seaweed jelly cube is another common sweetmeat of the area. Seaweed is washed several times until it becomes lighter in colour and is cooked in a little water. Lime and colouring are added to the syrupy mass obtained, which is then allowed to set for about an hour to a jelly like mass. The jelly is then cut into cubes.

Limiting Factors and Constraints

Over the last few decades, interest in the seaweed resources of Sri Lanka has surfaced periodically. There has been no continuing interest in the commercial exploitation of this resource, in spite of the marked growth in seaweed consumption and the seaweed-based industries worldwide. In Sri Lanka, the seaweed industry has always been a subsidiary of the fishing industry, with the collection and processing of seaweed being carried out by the fishermen to supplement their income. Unlike other countries of the region, the industry in Sri Lanka has so far not attracted adequate investment or interest from the industrial sector.

There are nearly 260 species of seaweeds growing in the coastal waters of Sri Lanka, of which nearly 20 species are of commercial interest (Durairatnam, 1961). Of these, only two species of Gracilaria are of any commercial importance at present. The under-utilization of the country's seaweed resources can be attributed to several limiting factors and constraints.

Non-availabillty of Resource Data

Information on Gracilaria spp is scanty. The lack of resource data has hampered the expansion of the industry, the collectors and exporters depending heavily on the resources from traditional collecting grounds. Exporters have often experienced difficulties in getting adequate supplies of dried seaweed to fulfil export orders. The unsettled conditions prevailing in the North and the East have also deprived exporters of access to the resources of the Trincomalee area, an area which used to provide 50-70 tonnes of dried seaweed annually. This has created a vacuum in the export market, a vacuum which could have been filled if alternative collecting grounds were available.

In a heavily export-oriented industry like the seaweed industry, uncertainties in the supply situation could restrict growth. The unavailability of resource data can thus be considered as one of the factors limiting the expansion of the seaweed industry in Sri Lanka.

Deficiencies and Difficulties in Collection and Marketing

Several deficiencies and difficulties in the collection and marketing of the economically important species of seaweeds of Sri Lanka can be identified. Gracilaria caters mainly to an export-oriented market, whereas Surgussum-based products have a fairly important local market. Imports of alginates, agar-agar, alginic acid substitutes like gelatine and related products in 1987 have been valued at Rs. 29 million.

When considering the past production figures for Gracilaria from the Eastern and North-western coasts, the potential production using traditional harvesting practices can be estimated at around 200 tonnes of dried seaweed. However, if other potential sites on the North-western and Northern coasts are considered, there is no doubt that there could be a significant increase in potential production. In the Kalpitiya area, Gracilaria is collected by waders, who often face difficulties due to the presence of poisonous sea snakes, sting rays and sharp edged shells of the razor shell oyster (Pinna). In the Mannar and Trincomalee areas, the collectors depend mainly on the seaweed cast ashore as only a very few beds have been located, and these are found mostly at depths ranging from 10-20 feet.

Production in the Kalpitiya area consists mainly of Gracilaria collected from shallow coastal beds. Collection is undertaken in response to demand for the dried seaweed from middlemen who supply the exporters. The collector fishermen cannot, therefore, depend on seaweeds for a regular income because the demand and market price are influenced by the variable export market. It is interesting to note that, at times, even the seaweed cast ashore is not collected by fishermen, because of difficulty in marketing it. Hence, it is important to devise a scheme for purchasing the product, independent of the middlemen, possibly through state intervention, in order to encourage the collector fishermen and to assure them a steady income.

As discussed earlier, most of the Gracilaria produced in Sri Lanka is exported in dried form without any further processing. Most countries exporting this seaweed have realised the benefits of exporting processed products rather than the dried seaweed. Export of the unprocessed product has two distinct disadvantages: the high costs of packaging and transport force down the price paid to collectors, and the raw material being processed elsewhere means that the benefits of gainful employment in manufacture are lost to Sri Lanka. A close look at the feasibility of further processing the seaweed locally is necessary, in order to improve the profitability and employment generation capacity of the industry.

Imports of Seaweed Products and Substitutes

During the period 1960-1970, several pilot-scale operations were undertaken to examine the feasibility of extracting alginic acid and agar from local seaweeds. The locally manufactured alginic acid and agar were found to be of international standard. Encouraged by the results of these trials, the Government decided to expand the industry, and action was taken to set up a network of collection centres and processing plants under the management of the District Development Councils. Interest in this direction was catalysed by the introduction by the Government of an import quota system for manufacturers.

A change in Government policy in the mid-1970s meant the removal of the import quota system, and the local market once again became open to imported seaweed products. This acted as a disincentive for investment in the local seaweed processing industry, and to date has resulted in Sri Lanka remaining as an exporter of dried seaweed. Attempts were also made in the 1970s to popularise the incorporation of dried powdered seaweed (Ulva spp) as a green supplement for poultry feed in place of imported alfalfa preparations. Laboratory studies were encouraging, with seaweed preparations showing promising results. Research and development activities on this subject were abandoned with the influx of cheaper feed material.

Agar is imported and marketed in Malaysia in four main forms: agar strips, bacteriological agar (powder), agar desserts (jellies) and flavoured powder mixes. The domestic marketing of agar entails a host of different intermediaries, each performing varied functions such as importing, wholesaling, .processing, distributing, merchandising, financing and retailing. Malaysia imported 172 tonnes of agar strips worth M$ 6.55 millions in 1988, which suggest that a large domestic market exists to support the production and processing of seaweed in the country. However, competition from other countries in the region may be a serious constraint for the development of Malaysian seaweed processing and agar production.

The study relied on secondary sources and field surveys during which interviews were carried out with importers, processors and dealers involved in the agar trade. Total trade volumes and values of agar imported for the last five years were obtained from the external trade statistics of the Department of Statistics. Preliminary surveys of agar products sold by retailers indicated that the main exporters of agar to

Malaysia were Korea, Japan and Singapore. Therefore the export figures of agar from these countries into Malaysia were obtained from the Japanese External Trade Organisation, and the Korean Trade Centre in Kuala Lumpur, and the Singapore Trade Statistics Department in Singapore.

Agar is imported and marketed in Malaysia in four main forms: agar strips, bacteriological agar (powder), agar desserts (jellies) and flavoured powder mixes. Five categories of intermediaries involved with the trade and marketing of the agar were identified: namely importer, wholesaler, distributor/packer, manufacturer and retailer. At least three from each category were interviewed. The surveys were carried out in Kuala Lumpur, Penang and Johore Bahru, which have been identified as the main centres for the trade and import of agar in the country.

Current Status of the Malaysian Seaweed Industry

Seaweed Resources of Malaysia

Traditionally, seaweeds have been utilised in Malaysia as food, animal feed and fertiliser and even in traditional medicine (Zaneveld, 1959; Burkill, 1966; Johnson, 1967; McHugh and Lanier, 1983). The documented use of local seaweeds was compiled by Phang (1984). Fourteen species of Agardhhiella, Corallopsis, Gelidiopsis, Gelidium Gr,acilaria, Grateloupia, Laurenciaa nd Padina were recorded as agar-producers, four species of Eucheuma as carrageenan-producers, and seven species of Colpomenia, Hormophysa, Hydroclathrus, Padi naand Sargassum as alginic acidproducers. Recent studies at the University of Malaya have added two more species of agarophytes, namely Gracilaria blodgettiia nd G. salicornia.

Research on Agarophytes and their Culture in Malaysia

The first study on agar content and quality of seaweeds in Malaysia was on Gracilaria by Burkill et al (1968). There was no further contribution to agar research until Doty & Fisher (1987) showed the potential for seaweed farming in Malaysia. The agar content of local species analysed ranged from 44-67% (Doty, et al 1983, Santos and Doty 1983). As a result of Doty's studies, a one-year pilot project on seaweed culture in Penang was undertaken (1983-84) by the Fisheries Research Institute at Glugor together with Agronomic Research Development and Production Inc., Honolulu, Hawaii. Further trials were reported by Faaaz Abdul Latiff (1986).

Production of Agarophytes and Agar in Malaysia

There is no known industrial production of agar in the country. Local fishermen collect species of Gracilaria and prepare the dessert "agar-agar" for home consumption or for sale. The first seaweed production company in Malaysia was set up in early 1989 based on the Gracilaria species known as G16. Due to unforeseen circumstances the crop was destroyed and plans for the continuation of the project are unknown. Gracilaria G16 had an agar yield of 17-19%, with gel strength greater than 610 g/cm^2.

The Agar Trade

In 1980, estimated world production of seaweeds for use in agar manufacture was 36,094 tonnes, of which about 18,088 tonnes was produced by Asian countries (McHugh and Lanier, 1983). Of the total amount produced in Asia, Korea accounted for 11,308 tonnes or 62.5% while Japan produced 4,OOO tonnes or 22.1%. The remainder came from the Philippines, Taiwan, India and Sri Lanka. It is estimated that the agar produced by Asian countries, mainly Japan, Korea, and to a lesser extent, Taiwan and India, was 3,574 tonnes. The primary consuming countries for agar are Japan, Korea and China.

Forms of Agar in Malaysia

Agar Strips

These are imported in bulk, mainly from Korea and Japan. The strips are repacked by packers or wholesalers into convenient sizes of 20 g or 35 g and then distributed directly or via retail outlets such as supermarkets and provision shops. The agar strips are bought by consumers for the preparation of jelly desserts, and the business enjoys peak sales during the many festivals in the country. Agar strips and agar powder are also imported,directly for utilisation by the manufacturers of jelly desserts.

Agar Desserts/jellies and Flavoured Powder Mixes

There are very few manufacturers of jelly desserts in the country, Socma being the main manufacturer. Some are imported in processed form from Japan, Indonesia and the Philippines. Instant agar powder mixes of various flavours are also sold. These come mainly from Thailand, and at least one brand is from Taiwan. They are sold in vacuum packs ranging from 10-85 g, or in paper boxes of 90 g.

Bacteriological Agar

Most of the bacteriological agar is sold in Kuala Lumpur to universities, research institutes and hospitals. The various brands available include "Difco", "Merck". "Oxoid", "Bitek", "Eiiken" and "High Media". They come mainly from the United States of America and United Kingdom. "Eiiken" comes from Japan, while "High Media" is the newest brand from India. The Indian agar is price competitive. One salesman estimates that the total market for bacteriological agar in the country is valued at M$ 2 million per annum.

Imports

The only official source for import figures is the external trade statistics published by the Malaysian Department of Statistics. One major disadvantage, however. is that statistics on agar imports do not appear separately but are lumped together under "seaweed products" (SITC code 292-981-O) or under "vegetable saps and extracts; pectic substitutes; pectinates and pectates; agar-agar and other mucilages" (SITC code 292-910-00). Hence, imports of agar can only be estimated by examining the statistics available for the two commodity groups.

The value of imported seaweed products increased from 1980 to 1987 as indicated in Table 1. Between 1984 and 1987, however, there was a decline in the value of imports due, perhaps, to the severe recession in the country. The main countries exporting agar to Malaysia are the People's Republic of China (which accounted for almost 37.0% of the total import value in 1987). Japan (18.1%). Korea (14.6%), Singapore (14.3%) and Taiwan (10.7%).

Marketing and Distribution Patterns

This section analyses the marketing of agar strips, agar powder and jelly agar in Malaysia. The marketing of agar is undertaken by a host of intermediaries such as importers, wholesalers, packers, agent distributors, processors/ manufacturers and retailers. All of these normally perform the three major functions of marketing, merchandising and financing simultaneously. In general, the marketing of agar follows the pattern of a few large importers selling direct to many wholesalers and/or agent distributors, who then service numerous small retailers such as provision shops, supermarkets, hotels and restaurants.

Agar Strip and Powder

Since practically all of the country's need for agar is imported, the role of importers in the marketing of agar strips is crucial. There are fewer than 20 companies, with various scales of operation, which have a monopoly over the trade. In a sense, therefore, the market structure of agar strips could be characterised as oligopolistic, with a small number of importers controlling a sizeable share of the agar market. More than half of the importers are found in Kuala Lumpur, while the remainder are in the other large cities such as Penang, Johore Bahru and Kuantan. Virtually all the importers also deal in other dried food provisions, and agar strip constitutes only a small proportion (less than 10%) of their total business volume. Three major types of agar importers could be identified, namely importer, importer-repacker and importer-repacker-wholesaler.

These three groups could be differentiated by the marketing activities they perform. The distribution channels and functions performed by the intermediaries in the marketing of agar strip is shown in Figure 1. It should be noted that the marketing and distribution of agar powder, another form of raw agar, is essentially similar to that of agar strip. Again, most if not all of the importers/ repackers/wholesalers of agar powder are found in large cities and major consumption centres such as Kuala Lumpur, Penang, Johore Bahru, Ipoh and Kuantan.

Like the importers, the wholesalers of agar strip are located mainly in large urban centres, and are involved in wholesaling other food provisions besides agar. They provide a crucial link between importers and retailers.

They also perform important facilitating functions such as selling, pricing, financing, grading and merchandising. A gross estimate suggests that more than 95% of the agar handled by wholesalers is sold to distributors/suppliers, manufacturers and large retailers, while only 2-5% went to small retailers and ultimate consumers.

Another marketing intermediary which is becoming increasingly important is the distributor/ supplier, otherwise known as the stockist. These provide a link between wholesalers and certain retailers, especially the small retailers scattered throughout the country.

At the retail level, the agar trade seems highly competitive as there are a large number of retailers involved. Agar retailing is generally widespread and agar is sold in most retail outlets; ranging from small village provision shops to large city supermarkets. The retailers provide the final link between sellers and consumers. Agar, however, accounted for a mere l-2% of the total volume handled by most retailers.

For the majority of agar traders (importers, wholesalers and retailers), agar constitutes only 5-10% of their total business. The actual volume of agar traded varies considerably from one dealer to another, depending on their scale of operation. A gross estimation indicates that the volume traded by each importer/wholesaler in a year, ranged from 3.0-10.0 tonnes valued at M$135,000-$450,000.

As for the retailers, it was found that their trade volume in agar is generally low, hardly exceeding 100 packets (20 g per packet) per month valued at M$ 160-$180. One large supermarket, however, reported that its sale of agar strip averages 600 packets a month, valued at M$ 960-$1000. Sales of agar powder are reported to be much lower, and retailers complained that it is not as saleable as agar strip. As expected, sales of both strip and powder increased by as much as 30-40% during festive seasons such as Hari Raya, Chinese New Year and Christmas.

Processed Agar (jelly desserts)

The manufacturing or processing of agar into jelly desserts is a relatively small business in Malaysia. Fewer than 10 companies are involved. Among the big names are Nutritional Products Sdn. Bhd. (Petaling Jaya) which is a subsidiary of Nestle, Foodpro Sdn. Bhd. (Kuala Lumpur), CPC (Malaysia) Sdn. Bhd. (Kuala Lumpur), Kinos Food Industries Sdn. Bhd. (Johore Bahru), and Fourseason Foodstuff Industries (M) Sdn. Bhd. (Penang). Except for Foodpro Sdn. Bhd., which is a subsidiary of an oil-refining company, the others are primarily food-based companies involved in the manufacturing of other food items.

Detailed business information could not easily be extracted from the manufacturers interviewed because of the high degree of secrecy maintained by most companies. However, some basic information on the manufacturing business of one particular company was obtained.

The company started to manufacture jelly around 1986 as part of its diversification programme. It is understood that the dessert accounts for about 20.0% of the company's overall business. The raw material required, agar powder, is imported mainly from Korea.

In 1988, the company imported approximately 6.0 tonnes of agar powder worth some M$ 150.000. The company sold about 2.7 million cups of jellies worth about M$ 1.0 million in 1988, an increase of about 10% over the previous year's sale of 2.4 million cups worth M$ 0.9 million.

The marketing pattern for the jelly desserts is somewhat different to that of agar strip or agar powder. A characteristic of the marketing of jelly desserts is the absence of wholesaling.

Instead, marketing is usually dominated by sole agent distributors and stockists/suppliers who provide the important link between the manufacturers and retailers. Most manufacturers undertake their own packaging and branding, while some like Foodpro Sdn. Bhd. also do contract packaging for retailers such as the Seven Eleven Company. The retailing of jelly desserts is done mostly by supermarkets, mini-markets and provision shops selling food items. These outlets are dispersed throughout the country.

As with agar strip retailing, jelly desserts constitute only one of the many items sold, and usually account for less than 10% of the total volume.

Bacteriological Agar

Marketing of bacteriological agar is more simple than that of agar strip or processed agar jellies. Essentially, it revolves around the importer-cum-distributor who imports directly from overseas manufacturers, in particular from the U.S.A., United Kingdom/Europe.

Field surveys identified companies selling scientific equipment and supplies as the main businesses involved in the bacteriological agar trade. Importing small quantities at a time, most of them reported that bacteriological agar constitutes only a fraction (1-2%) of their total business.

A unique feature of the bacteriological agar trade is that neither wholesaling nor retailing is involved in its marketing and distribution. The product is sold directly by the importer/ distributor to the consumers. The main consumers are hospitals, universities and research institutes. In some instances, these institutions import bacteriological agar direct.

Demand and Prices

International demand for agar remains strong as evidenced by the continuing high priëes and the ease with which major exporting countries like.China and Korea (McHugh and Lanier, 1983) dispose of the product. The market price of agar seaweed is higher than those for other colloidbearing seaweeds.

In Malaysia, demand for agar has been dictated primarily by its major utilisation, as food for human consumption. Unlike more essential food items such as rice, fish, meat and vegetables, agar is considered non-essential and is eaten only as a desert. Hence, demand for agar is very income and price elastic, meaning that a slight change in the price of agar and consumers' income would bring about a remarkable change in the demand.

Besides price and income, demand for agar is also determined by consumer preference, availability and price of other colloid substitutes (e.g. carrageenans), eating habits and nutritional considerations. One major advantage is agar's acceptability to all religious and cultural groups among multi-racial Malaysian communities.

Statistics on the domestic demand for agar and its per capita consumption are not, how-ever, available. The market price of agar is reflected, to some extent, in its import value (cif). From the statistics available on imports from Korea, it was found that the ex-vessel prices of agar strip have been on the increase over the last five to six years, averaging M$37.95 per kilogram in 1988.

This represents a remarkable increase from the 1982 price estimated at M$24.70 per kilogram. On average, the import value of agar strip increased at an annual rate of about 10.0%. The rising prices of imported agar can be construed as reflecting the continuing strong domestic demand.

At wholesale and retail levels, there can be wide fluctuations in the price of agar strips depending on the individual trader's profit margin and the quality of the agar as measured by its gel-strength. Laboratory analyses were undertaken to determine the relationship between gel strength and retail prices. As expected, it was found that the higher the gel-strength of the agar strip. the higher its retail price.

On average, the high-quality agar strip, originating mainly from Korea, is retailed at M$ 1.80- M$ 1.90 per 20 g packet. In comparison, the retail price for low- quality agar strip, supplied mainly from China, averages M$ 1.20- M$ 1.30 per 20 g packet. The retail price of agar strip of the same quality may differ by as much as 10% from one outlet to another, depending on the individual retailer's profit margin.

The prices of jelly desserts were more difficult to estimate, as they are sold in different types of containers with different units of measurement. The most common packaging is in packets containing 20 cups of jellies, each individually packed with 19 g content. A packet of agar jellies is usually retailed at M$1.00-M$1.30, with slight variations from one retailer to another. A well-known brand, Square Cut Jelly, is sold at M$0.60 per cup, and the content per cup is 130 g. In the case of bacteriological agar, it was found that the consumer price varies in the region of M$250-M$450 per kilogram depending on the grade. For example, Bacto-Agar, a supposedly high-grade bacteriological agar is sold at M$ 418 per kilogram, while a lower grade called Bitek-Agar fetches about M$352 per kilogram. The purchase price of bacteriological agar was in the region of M$ 200- M$ 350 per kilogram, depending on the grade. In general, the purchase price had increased by as much as 10-20% since 1986.

Marketing Margins

The marketing margin or price spread for agar in this study is analysed in terms of dollars per kilogram and as percentage of retail price. In an efficient marketing system. the marketing margins can be equated with processing/manufacturing costs, transportation costs, handling charges and the normal profits of market intermediaries.

At the wholesale level, the marketing margin and wholesale price account for 33.3% and 55.5% respectively of the retail price. At the retail level, the marketing margin is 11.1% of the retail price, all of which constitutes gross profit margin to the retailers.

Analysis of the marketing margin indicates that the importer's share of the retail price is M$0.60 per 20 g packet or 33.3%. Based on this information, the importer's share of the retail price appears relatively low. This does not necessarily mean that the marketing system is inefficient.

Potential for Development and Constraints

Malaysia imported nearly 172 tonnes of agar strip worth some M$6.55 million in 1988. This means that a sufficiently large domestic market exists to support the production and processing of seaweed, especially Gracilaria, in the country. Moreover, if this species could be successfully cultivated in Malaysia, substantial savings in foreign exchange would be attained through importsubstitution.

Malaysia is blessed with several natural factors such as an abundance of sunshine and the absence of destructive typhoons. Most of the coastal areas are relatively unpolluted from sources such as silting, agricultural, agro-industrial and human wastes.

Aquaculture in general has not been intensively developed, and at present offers no strong competition to seaweed culture. The political stability of the country, together with its well developed infrastructure and good shipping and communication network, pose few problems for the marketing of the product. A substantial domestic market for the processed agar, as well as good regional (ASEAN, Japan and Korea) and possibly international markets (U.S.A.. E.E.C., etc.), is also envisaged. Malaysia has good international links with the Muslim world and agar is the accepted substitute for gelatine in the food, cosmetic and pharmaceutical industries.

Notwithstanding the bright prospects there exist a number of constraints and problems that need to be overcome if a seaweed industry is to be developed. Heading this list is the inadequacy of technical capability and expertise.

The technology available for seaweed culture in Malaysia is far from satisfactory and it may take several years of concentrated and well-designed effort to make it successful enough to compete with other seaweed producing countries in the region.

A constant supply of consistent quality is important if seaweed produced in Malaysia is to be competitive with that of other producers. Moreover, while Gracilaria cylindrica was shown to produce a good food agar, it cannot be used for more demanding and higher valued uses such as bacteriological agar. The search for high-quality agar producing species must therefore continue. Stock improvement studies should also be conducted.

In addition to technological considerations, the potential development of the seaweed industry in Malaysia may be affected by market and economic constraints. While the domestic market may be able to support a small number of agar-producing enterprises, the full-scale development of the agar industry will very much depend on a reliable regional and international market. Unfortunately, potential producers (particularly those in developing countries like Malaysia) wishing to enter the international seaweed trade would be faced with stiff competition from the more established producers.

CONCLUSION

Almost all of the country's agar is currently imported, mainly from Korea. The bulk is utilised in Malaysia in the form of agar strips used mainly for jelly desserts. The other forms include agar powder and bacteriological agar for scientific purposes.

The domestic marketing of agar entails a host of different intermediaries, each performing varied functions such as importing, wholesaling, processing, distributing, merchandising, financing and retailing. In essence, the marketing of agar follows the general pattern of a few large importers selling to a relatively large number of wholesalers and/or agent distributors and stockists, who then service numerous small retailers and other retail outlets such as supermarkets, restaurants and hotels. Domestic demand remains strong as suggested by the continuing high imports and cif prices of agar.

However, demand was adversely affected by the 1984-86 economic recession which had the effect of reducing imports, particularly from Japan. Furthermore, the fall in imports from Japan since 1986 can be attributed to the rising value of the yen.

Demand and market potential for agar could not be analysed in detail, owing to the absence of detailed information. However, it is unlikely that demand for agar would increase substantially in the future since agar is a non-essential food item. Demand for food agar, however, is very price and income elastic. It is timely, perhaps, for Malaysia to develop seaweed production and processing.

However, before venturing into seaweed culture and processing on a large scale, there is a need to undertake in-depth market evaluation for different species of seaweeds, both at the domestic and international levels. New markets should be explored so that new seaweed-producing countries like Malaysia would not be too vulnerable to wide price fluctuation and uncertainties.

The domestic market should be expanded to reduce over-dependence on the foreign market. There is also a need for more research on new uses for the various species of seaweeds. To ensure a constant supply of raw seaweeds for any processing plant that might be set up, Malaysia could act as a collecting centre for neighbouring countries in the ASEAN region. Once processed, the product could then be sold back to these countries.

INTERNATIONAL TRADE IN AGAR

Trade statistics from 1980-87 of imported/exported agar from the BOBP countries are presented in this paper. Thailand, Malaysia and Indonesia account for most of the imports in the region. Export of agar is recorded from four BOBP countries. However, re-export and misclassification of items may interfere with sound interpretaion of the statistics.

It is the intention, here, to look at the international trade in agar, making use of available statistics. Before examining the data it is prudent to sound a cautionary note. Trade statistics are ultimately only as good as the person who completes the customs form or the one who interprets it. Items may be misclassified. So-called 'Danish agar', for example, which appears in European trade statistics, is not agar at all but furcellaran, a carrageenan-type gum obtained from the seaweed Furcellaria. Figures available for imports of agar into the European Community indicate that 301 tonnes from the Philippines entered the UK in 1986 (Nimeke, 1986), a figure cited recently elsewhere (Anon, 1988), although incorrectly attributed to 1987.

This, too, according to sources in the United Kingdom and Philippines (personal communication), is spurious and presumably refers either to carrageenan or the seaweed from which it is obtained.

For some countries, agar is not separated from other commodities. This is the case for Malaysia, for example, where it is included under the heading, 'Vegetable saps and extracts, pectic substances, pectinates, agar and other mucilages'. In such a situation, imports of agar can only be estimated by examining other countries' exports. Even this, however, may not be accurate or reliable.

Thus, figures for imports of agar into Indonesia from Japan, when extracted from Indonesian import and Japanese export statistics, respectively, are in poor agreement although both list agar separately.

Japan Exports and Imports

Notwithstanding these comments, it is hoped that the trade statistics presented go some way towards representing the true picture with regard to international trade in agar within the BOBP region.

Thailand, Malaysia and Indonesia account for most of the imports of agar within the region. Imports into Thailand, valued at Bht 112.9 million in 1987, come mainly from Japan, although there have been increasing amounts from Chile in recent years.

Figures for Malaysian imports have been fairly consistent during the present decade, at around 250 tonnes per annum, except for 1983 when a sharp increase to over 500 tonnes was recorded. Singapore, it should be noted, is not a producer of agar, but rather, re-exports imported material.

Indonesia imports agar from a variety of sources, with Asian and South American countries contributing significant amounts. Figures for Hong Kong and Malaysia, like those of Singapore, refer to re-exports. Total imports for 1987 were valued at US $526,000. Imports in 1983, again, appear to have been higher than normal.

Indian demand for agar is met primarily by domestic production (Coppen, 1989) but small amounts of higher grade bacteriological/pharmaceutical agar are imported. Imports of agar into Sri Lanka and Bangladesh are also low, of the order of a few tonnes per annum or less. There is not much local production either.

Exports of agar are recorded for four BOBP countries, although for Thailand and Malaysia, at least, these probably represent re-exports.

The main destination for Thai agar has been the USA. For Indonesia, most of the 9 tonnes exported in 1988 went to Singapore.

8

Management of Natural Resources of Tropical Agarophytes

INTRODUCTION

The production of agar-producing seaweeds comes from three sources, namely gathering of drift materials, direct harvest from natural stocks and cultivation. At present, the bulk of the seaweeds used for agar manufacture still comes from harvested natural stocks. There are no available and accurate data on the contribution of agarophytes produced through culture, but judging from the genera presently produced in different countries, which were imported by Japan in 1984 (Armisen and Galatas 1987). about 50% of the raw seaweeds presently processed into agar still come from natural stocks. The genera presently utilized in the international market for agar production are Grucilaria, Gelidium, Pterocladia, Gelidiella, Anfheltia and Ceramium.

Approximately 6,683 tonnes of agar were produced by 14 countries in 1984 (Armisen and Galatas 1987). Of this, about 48% was processed from Gelidium and Pterocludia with the remaining 57% from "other seaweeds" which undoubtedly include genera such as Gelidiella and Gracilaria. As far as we know today, only the genus Gracilaria is presently produced in commercial quantities through cultivation. In the first half of the 1980s, Chile, Brazil, Taiwan and Philippines were the main suppliers of cultured Gracilaria to Japan. Recently, Vietnam, Indonesia, Thailand, Hawaii and others have applied culture techniques in Gracilaria production. At present, however, a large proportion of the Gracilaria produced still comes from local stocks. Gelidium and Pterocladia, utilized mainly for the manufacture of bacteriological agar and agarose, come so far as we are aware, from the gathering of natural stocks. Their commercial production through farming seems to be still a long way from realization. Thus, the production of most of these agarophytes is expected to be dependent on their natural stocks for a long time to come.

IMPORTANT CONSIDERATIONS IN THE UTILIZATION OF NATURAL STOCKS AS A SOURCE OF BIOMASS

The production of agarophytes from natural stocks is very much influenced both by seasonal factors and by harvest pressures exerted on them during the preceding cropping season. Because their growth cycles are greatly influenced by environmental conditions and by man's exploitive activities, their production is unreliable. They are prone to over-exploitation, and the need to manage and conserve their stocks is of prime importance in order to sustain or further enhance their productivity and prevent over-exploitation.

The design of a sound management scheme for the natural stocks of commercial agarophytes depends primarily on the availability of information on the various aspects of their biology, such as reproduction and growth cycle, growth rates, their regeneration and recruitment capacities, their productivity and the influence of environmental factors on the biomass production potential of the stocks. This information is necessary to formulate guidelines for managing the natural stock of the target species. The information can provide answers to questions such as where the species is abundant, how much to harvest per unit area, when to harvest, how many times (cropping intervals) the stocks can be harvested in one season and what kind of harvest method is best for the species.

Gathering such information on the species to be managed requires basic skills in field sampling and data gathering. Under management. the production of stocks can be forecast with confidence. This information is most important for quoting in contracts which may be entered into by the farmer, fisherman.

or exporter. It is therefore of prime importance that any plan to exploit natural seaweed stocks be preceded by intensive biological studies to determine their seasonality in biomass production, reproduction, regeneration and recruitment. This information is necessary for determining the best possible time to harvest, and the amount of stocks harvestible, without diminishing their production capacities.

REQUISITES TO THE RATIONAL EXPLOITATION OF NATURAL STOCKS

In a "free-for-all" situation, there is always a natural tendency for resource users to over-exploit resources, especially where there is a prevailing demand for the produce. The literature is replete with records of resource losses due to over exploitation. Thus, it is necessary that the commercial exploitation of natural seaweed stocks, or any resource, be preceded by biological studies which shall be the basis for their management.

Inventory and Assessment of Stocks

The inventory and assessment of stocks are initial studies which should be done in areas where the exploitation of stocks has not yet started. This will identify the species potentially available for development, where and when these species are abundant, how much biomass is available for harvest and the behaviour and responses of the stocks to certain degrees of exploitation. The need to know the true identities of the different species is very important because unlike other resources, e.g., fish and crustaceans, the kinds and quality of agar vary from species to species.

Thus, it is very important that the taxonomy of the species comprising the stocks is known. In addition, the quality of the hydrocolloid (agar) should be defined, characterized or evaluated as the price of the produce is determined by the quality of its extractable agar. In the world agar market, the name of the species and, in the case of Gracilaria, information on the source of the dried raw material, .are important because they reflect the differences in the properties of the agars. Gracilaria from

Chile, for instance, fetches a higher price than that from the Philippines because it contains higher quality agar. Thus, the names applied to these produce serve as a basis for their pricing. The price is generally based on the moisture content and the purity and quality (gel strength, melting and gelling temperatures, viscosity and amount of sulfate group) of the agar.

Information on the abundance and distribution of the resource in space and time may be gathered through biomass samplings of the stocks. The application of available methodologies may differ slightly, depending on the behaviour of the resource. However, the transect quadrat method is widely used especially in situations where stocks are not homogeneously distributed in space. The size of the bed is first delineated and permanent transects are marked. The orientation of the transects is generally related to certain ecological gradients, such as depth and wave exposure. Biomass samplings are generally done on a monthly basis along the transects; the size of the quadrat varies from 0.25-1.0 m^2. The number of quadrats to be sampled along the transects may also vary, depending on the size and homogeneity of the bed, the time and effort required but most important is the amount and type of data for statistical analysis i.e. the more the samples that are gathered, the more reliable the data will be.

The amount of loose (drift) biomass should also be monitored to derive reliable data on the total biomass production of the bed.

Data on biomass production recorded for a period of one year will indicate the annual productivity and seasonality in production of the stocks. Additional data from following years will make the information on the stocks more reliable as a basis for management. In most stock assessment studies where the target species or group of species and their distribution are known, the main concern is the determination of production data.

The size of the bed must be known so that the potential total production of stocks can be determined. This information is vital in determining how much of the stocks should be harvested without unduly diminishing their productivity. Using the production data from the samples, total production of the bed may be projected/calculated by multiplying the biomass data (g/m2) by the size of the bed. The accuracy of the method is much improved if regular and repeated samplings of the stocks are carried out.

Additional information on production in the area can be gained through interviews with market vendors and seaweed gatherers. Initial interviews may be done in local open markets where seaweeds are sold. Seaweed vendors are a good source of information on the kinds of species, the amounts they sell, the sources of seaweed stocks, suppliers and the approximate number of gatherers. Seaweed gatherers are good sources of primary data on production. They can easily be identified by inquiring from local officials in the area. Data on gathering sites, number of gatherers, the gathering season and output per unit effort may be acquired from this source. An estimate of local production can then be made, and counter-checked with the data on potential biomass production of the beds or collecting area.

The seasonal variation in the reproductive/fertility states of the stocks should also be known. The fertility of the stock may be determined by randomly collecting 50 or more thalli and determining the number which are vegetative and fertile. This is usually expressed as % fertility. This information is relevant in the scheduling of harvesting/cropping periods. The recruitment capacities of stocks is generally influenced by their states of fertility, especially for those species where recruitment is largely dependent'on the production of reproductive cells (spores). Cropping or harvesting should be scheduled some time before or after the peak of fertility of the stocks, in order not to interfere unduly with the recruitment process. This, however, may not be relevant to species where production is based mainly on vegetative means (cuttings, fragments). Some stocks of Gracilaria, for instance, have been reported to be purely vegetative (Rueness et al. 1987). Pond cultured Gruciluriu have been observed to be purely vegetative the whole year round in Northern Philippines (personal communication). In cases where recruitment is primarily based on vegetative means, a certain amount of the seedstock is retained in the bed for the next season's croppings. This amount may be equivalent to the amount of biomass produced in the bed during its lowest production period.

In addition to these basic biological studies, harvesting experiments should be carried out to determine the production capacities and the effects of different harvest pressures and methods on the regeneration of the stocks. The information on the capability of the stocks to regenerate to their former level of production after trial cropping/harvesting shall be the basis for determining the harvest schedules during the cropping season.

DEVELOPMENT AND APPLICATION OF A MANAGEMENT SCHEME FOR THE STOCKS

The application of a management scheme to natural stocks has been shown to significantly improve the total production of the beds. In Chile, the annual production of Gracilaria in Lenga Cove located in San Vicente Bay increased from 80 tonnes to 6OO tonnes after the application of a management programme (Poblete and Inostroza. 1987). It is apparent from their studies that the strict application of a management scheme had improved the production ecology of the bed, resulting in an increase of seven and a half times its normal production.

In general the formulation and application of a management scheme for the natural stock of seaweeds may follow the following steps. Variation from this scheme may be necessary to suit certain biological characteristics of the species concerned. The basic considerations are:

(a) Seasonal changes in the annual productivity of the stocks. Data on changes in the annual productivity of the stocks are derived from the monthly biomass measurements done on the bed. They provide information on the total amount of biomass available for cropping, and the season when cropping/harvesting may be done. In addition, the seasonal variation in the amount of agar contained in the crop is also considered in the scheduling of the harvest. A high quality crop is obtained when harvesting is done during periods where the agar content is high.

(b) Determination of the amount of biomass to be left after cropping. The amount of biomass left after the first cropping is very important in determining the amount of biomass available for the next cropping season. As a "rule of thumb," the minimum amount to be left in the bed to serve as "seeds" for the next cropping season should not be less than the lowest biomass recorded during the year. The amount of biomass available for harvest will be the difference between the total amount available in the bed during its peak production period and the minimum biomass during its period of depressed growth.

(c) Determination of the harvest schedule. How often should the harvest be done during the peak production season? The harvest regimes can be determined from information on the regenerative capacity of the stocks, e.g. from the results of studies on different trial harvest pressure. The schedule of subsequent harvests is determined by the period within which the stock can recover its original biomass after the first harvest.

(d) Control of the amount of biomass to be harvested. The amount of biomass to be harvested during each harvest regime should not exceed the amount of biomass available for cropping. It is, therefore, necessary that the number of fishermen should be limited, and that the amount of crop each is allowed to harvest must not exceed his share of the biomass available for cropping during each of the cropping periods.

(e) Protection of the recruitment and regeneration processes. This is a very important consideration because the continuity of the stocks depend on these processes. The approach may differ depending on the biological characteristics of the species. For stocks which depend on the regeneration of biomass through vegetative structures such as holdfasts, cuttings or fragments, and underground thalli, protection may be ensured by the application of harvest methods which cause the minimum harm to these regenerative structures. For stocks whose recruitment and regeneration processes depend both on reproductive cells and vegetative structures, the timing of the harvest so as not to impede or adversely affect recruitment and the use of harvest methods which have minimal destructive effects on the regenerative structures are important considerations. Thus, harvesting should not be done during peak fertility of the stocks.

(f) Socio-economic considerations in managing the resources. The fishermen/seaweed gatherers should be organized into production groups such as cooperatives. Only bonafide members have the right of access to the resource, and the amount each member is allowed to crop is determined by his fair share of the biomass available for cropping. A legal basis for enforcing these regulations can easily be achieved through rules promulgated by the local government, or by the cooperative's established rules/regulations. Peer pressure among the members can also be a strong factor. Management of the co-operative is done by selected members. A portion of the members' earnings is channelled back to the co-operative for management support. Technical assistance should be extended to the cooperative by the government agencies concerned.

The genus Gracilaria is represented by 32 species in the Indian region. of which 31 are found in different parts of the Bay of Bengal coast. In respect of species diversity and biomass, Gruciluriu is second only to Surgussum of the Phaeophyta. Another noteworthy feature of Gruciluriu is the diversity of the species on the Indian coast which is not matched anywhere else in the world. It may, therefore, be stated without hesitation that Gracilaria is the most economically important genus on the Indian (Bay of Bengal) coast.

Two species of Gruciluriu, G. corticutua nd G. folifera. are found more or less throughout the Indian coasts wherever the shore is rocky. On the other hand, a few economically important species like G. edulis, G.v errucosa, G. fergusonii, G. bursa-pastoriasn d G. textoriih ave restricted distribution. G. eucheumoides is found only in the Nicobar group of Islands in the Bay of Bengal. Estimates of Gruciluriu resources on the Indian coasts have been made by several scientists using diverse methods.

Earlier estimates by Koshy & John (1948), Chacko & Malu Pillai (1958), Thivy (1960) and Desai (1967) gave no details of the methods of survey, and were restricted to small stretches of the coast. Varma and Krishna Rao (1962) and Umamaheswara Rao (1968) described the methods they used, although their surveys were confined to small areas in the Gulf of Mannar and Palk Bay sides of Mandapam. Harvest figures, based on landings for Pamban, Periapatnam and Kilakkarai during 1966 to 1968, were given by Umamaheswara Rao (1968).

A study of seaweed drift on the Indian coast was reported by Krishnamurthy et al (1967) and by Krishnamurthy (1969). These drifts are particularly heavy in some parts of the Indian coast and may be used locally for manure and for bio-gas production. This survey indicated potential areas for more intensive surveys.

The more scientific and, therefore, more reliable survey made by a team of scientists from the Central Salt and Marine Chemicals Research Institute, Bhavnagar, in collaboration with other institutions like the Central Marine Fisheries Research Institute and the Departments of Fisheries of various State Governments, gave data that can be taken as the basis of future estimates. The survey, begun on the Tamil Nadu coast, was followed by a similar survey on the coast of Andhra Pradesh.

A perusal of this table shows that the total Gracilaria resources of the east coast of India assessed so far can sustain a production of about 75 tonnes of agar per annum, assuming that all of the Gracilaria can be harvested and utilized. In practice, however, it is neither feasible nor desirable to harvest the entire quantity available. Assuming that 50% of the standing crop can be harvested, it follows that only about 37.5 tonnes of agar can be produced. This is totally inadequate and will not meet Indian demand for agar.

The agarophyte resources of India have been over-exploited and the biomass of these algae has been diminishing over the years. The estimates considered above are 15 to 20 years old and may not have much significance now. A renewed estimate, adopting a scientific method of survey and annual monitoring of the resources using remote sensing techniques, is essential to keep a watch over these resources.

Experimental work on repeated harvesting of a standing crop of Gelidiella acerosa (Joshi & Chauhan, 1985) has shown that adopting a sensible harvesting programme can ensure regrowth of the standing crop. Similar studies on Grucilaria would throw light on methods for conserving these resources. Studies on the repeated harvesting of cultivated Grucilaria edulis have shown that an interval of about six months between harvests will ensure continued good growth of the crop.

Indonesia's seaweeds of commercial importance are the agarophytes Gracilaria spp., Gelidium spp., Gelidiella spp. and Pterocladia spp., and the carrageenophytes Eucheuma spp. and Hypnea spp. Until 1983, almost all of the seaweed products manufactured in Indonesia were derived from the harvesting of wild seaweed stocks. These are neither adequate in volume and reliability, nor competitive in cost. There are also problems with continuity of supply and in controlling the quality and quantity of the products.

Commercial seaweed farming began in Indonesia on a small scale in 1979, using Eucheuma spinosum. After more intensive study and development, further E. spinosum farming started in 1983. Farming of E. cottonii (E. averizii)b egan in 1984, in cooperation with the Genu-Kopenhagen pectin factory.

Indonesian Agarophytes

An intensive study of Indonesian agarophytes was started in 1985. It involved collecting local species and analysing their quality; studying biological and environmental aspects in support of farming experiments; studying farming techniques and methods for experimental farm and farming development; and studying agar processing methods and the development of an agar processing industry.

Variation in the quality of some Indonesian Gracilaria

Indonesia's Gracilaria comprise three main species, namely G. lichenoides, G. gigas and G. verrucosa.

Gracilaria are used in the manufacture of various products which include human food, animal fodder, fertilizers and soil conditioners, and as raw materials for the chemical and pharmaceutical industries. The latter two are significant because they have a market worth of over US $300 million annually (HAI., 1986). Thailand exports seaweed (mostly Gradaria) totalling between 20 and 200 tonnes/year (dry weight), and valued at 4-10 million baht (HAI.. 1986) At present, eight endemic agarophytes are found in Thailand (Abbott, 1987), four species belonging to the genus Gracilaria (C. tenuistipitatav ar liui, G. firma, C. irregularis and C. salicornia and four to the genus Polycavernosa (P. fisheri, P. changii, P. fastigiata and P. percurrrens). The major producing areas are the southern region of Songkhla lake and the shallow mudflats in Pattani Bay (Tachanoravong. 1989).

The purpose of this study was to investigate the natural growth of endemic agarophytes in southwestern Thailand (Krabi, Trang and Satul provinces). The results can be used as basic information for natural production management and for future studies of culture techniques.

Method

The investigation was conducted between November 1987 and April 1988 in the coastal areas of Krabi, Trang and Satul provinces. A pre-survey in the three provinces was done in November 1987, in order to study basic information on natural production of Gracifaria and Polycavernosa. The investigation was carried out at 1 1 stations in Krabi province. six stations in Trang province and 10 stations in Satul province. Every survey station was investigated twice during the period of study.

Water and air temperatures were recorded by mercury thermometer, and water salinity was measured by hand refractometer.

The natural growth of seaweed which attached to fin-fish net cages was recorded. The growth period and seasonal fluctuations were observed. The total amount of endemic agarophytes in the fin-fish net cages was weighed. An estimate of production for the harvesting season was made as follows:

1. The cages with attached Polycavernosa were taken out and weighed by direct balance (wet weight). (The size of fin-fish net cages is around 5 x 5 m and 2 m in depth).
2. The frequency of cleaning the net cages was recorded. The weight of Polycavernosa removed when cleaning the net cages each time was estimated and recorded.
3. Total amount of Polycavernosa which could possibly be harvested in a year was estimated.

The genus Capsicum includes approximately 20 to 25 species, all originated from the American continent. This estimate is, however, only approximate and it is possible that new species are discovered when germplasm collection expeditions in the wild are undertaken. Only five species of Capsicum are largely cultivated (C. annuum, C. baccatum, C. chinense, C. frutescens and C. pubescens) while the great majority are not yet domesticated. There are many taxonomic proposals for the wild species of the genus, with considerable controversies to be addressed (Prince et al., 1995). The taxonomic and genetic variability studies of Capsicum species native to Brazil have been based, mainly, on morphological characters, interespecific crossings and fertility of interspecific hybrids. Little is known about the Brazilian wild species, their genetic variability and phylogenetic relationships. The biological knowledge of this germplasm is important for the understanding of the genus in its totality. This knowledge is also fundamental for the development of appropriate strategies of germplasm collection and conservation. Additionally, it is possible that the wild germplasm harbors genes of agronomic importance, such as those related to disease resistance, yield and fruit quality.

Recently, a collection expedition to the Atlantic forest in Eastern Brazil and Roraima state sampled wild populations of Capsicum in areas prone to high anthropic pressure. The main objective of the study described here was to analyze the genetic variability and phylogenetic relationships at the DNA level of a sample of the new accessions in comparison with wild and cultivated Capsicum genotypes maintained in a germplasm collection.

Material and Methods

Genetic Material: Biological samples of 92 accessions of wild and cultivated Capsicum species were used for DNA extraction, including samples of 26 new wild Capsicum accessions recently collected in the Atlantic Forest and Roraima state. Part of the accessions were analyzed with morphological descriptors and tentatively classified at the species level. Additionally, four species were used as outgroups in the study: Atropa belladona, Lycopersicon esculentum, Nicotiana tabacum and Physalis spp.

DNA polymorphism analysis with RAPD markers: The genomic DNA was isolated from young leaves according to Doyle and Doyle (1987). The PCR conditions followed the protocol described by Ferreira e Grattapaglia (1998). Each 10 μl of reaction contained: 10 mM Tris-HCl pH 8,3, 1,5 mM MgCl2, 1,0 μg/μl of BSA, 0,2 mM of each dNTP, 0,4 μM primer, 7,5 ng of genomic DNA and 1 unity of Taq polymerase.

Phenetic Analysis: Random amplified polymorphic DNA data were used to construct a binary matrix based on the presence or absence of the amplified fragment after separation of PCR products in electrophoresis gels. A genetic similarity analysis among accessions was performed using the Jaccard coefficient, and the resultant similarity matrix was used to group genotypes through UPGMA analysis (Sneath and Sokal, 1973). Dendrograms were produced using the NTSYS-pc 2.0 package for numerical taxonomy (Rohlf, 1992)

Cladistic Analysis: The character state was the fragment being either present in an accession (coded as 1) or absent (coded as 0). The final data set, therefore, consisted of a binary data matrix of the presence and absence of restriction fragments. Phylogenetic analysis was performed using PAUP 3.1.1 (Phylogenetic Analysis Using

Parsimony; Swofford, 1993) by unweighted maximum parsimony, considering each character as equally weighted and unordered. For tree searches Tree Bisection-Reconnection (TBR) branch swapping was selected with ACCTRAN (accelerated transformation) optimization. PAUP was also used for computing consensus trees (strict, semi-strict and 50% majority rule), performing bootstrap analysis to assess the degree of support for each branch and computing character statistics such as consistency index (CI) and retention index (RI).

Results and Discussion

Seventeen highly informative RAPD primers (Operon® primers AA1, AB8, AB9, AB11, AB17, C16, F13, O4, O7, O11, R1, R14, Y18, W4, W13, W19 e Z6) selected in previous experiments were used in the evaluation of the 96 accessions. A total of 277 polymorphic RAPD markers were used to analyze the 92 wild and cultivated Capsicum accessions and the four outgroup species. The phenetic and cladistic analyses (Figures 1 and 2) resulted both in dendrograms that are in general, compatible with the known Capsicum phylogeny (Prince et al., 1995). The cladistic analysis resulted in 20100 trees with maximum parsimony with length of 703 steps and CI=0,393 and RI=0,890. The clades were well defined and with good statistical support as indicated by the indices (CI e RI) and by the bootstrap values described in the tree.

The accessions were divided into two main groups: one group containing the cultivated, semi-domesticated and wild species collected in the Atlantic Forest and Roraima state and a second group with the outgroup species clustered.

The first group was divided in two subgroups. In the first subgroup, the C. annuum, C. frutescens e C. chinense accessions presented high relative similarity, while the C. baccatum e C.

baccatum var. praetermissum accessions constituted another subgroup. Three accessions (94,95 e 96), recently collected in Roraima state and classified based on morphological and botanical descriptors as C. chinense, grouped with accessions of C. chinenese species (Figures 1 e 2). The second subgroup included the remaining accessions collected in the Brazilian Atlantic Forest, which clustered with samples of other wild accessions such as C. buforum, C. villosum, C. dusenii, C. parvifolium, C. campilopodium, C. fexuosum e C. chacoense. The phenetic analysis suggested a subgroup composed of C. flexuosum e C. chacoense accessions, what was not corroborated by the cladistic analysis. Similarly, C. villosum and C. buforum accessions formed a subgroup that in the phenetic analysis was genetically similar to the newly collected sp1, sp2 e sp3, but that clustered with sp4, sp6, sp7, sp8 e sp9 based on the cladistic analysis. However, both analysis indicated that the accessions sp1, sp2 e sp3 were genetically similar. The data obtained in this study allowed the estimate of genetic diversity and the establishment of phylogenetic relationships of the germplasm analyzed. Complementary studies of cpDNA sequence divergence are currently underway to further investigate the existence of new Capsicum species within the recently collected accessions.

9

GERMPLASM COLLECTION OF CAPSICUM

INTRODUCTION

The cultivation of sweet pepper and other Capsicum spp. species are of increasing relevance for the Brazilian agriculture. Annually, ca. 13,000 ha are cultivated yielding approximately 280,000 tons of fruits for fresh consumption and processing as sauces, preserves and other preparation forms.

The agribusiness activities involving Capsicum demand new cultivars, particularly genotypes for processing with higher yield and quality, and other important traits such as resistance to the most important diseases and insects. Breeding new cultivars is based on genetic variability, and for that a well-organized and characterized germplasm bank is essential. Embrapa Hortaliças (CNPH) has begin a Capsicum germplasm bank 23 years ago, in cooperation with other Brazilian and international institutions. In 1998, a research project financed by the World Bank (Prodetab) was established to enrich, characterize and maintain the germplasm bank, as well as to organize a database to supply relevant information to the Capsicum breeding program.

The Embrapa Hortaliças Capsicum germplasm bank was established in 1980 by germplasm exchange with other countries and domestic collect expeditions. Presently, the germplasm bank has 1,220 entries of open pollinization cultivars, hybrids, populations of cultivated types, lines and wild types of C. annuum, C. baccatum, C. chinense, C. frutescens and C. pubescens.

The Capsicum research project developed by Embrapa Hortaliças has a multidisciplinary and interinstitutional team involving germplasm collection; multiplication; morphological, cytogenetic and molecular characterization; conservation; documentation; evaluation of disease resistance; and plant breeding. Presently, the breeding program is focused on the demands of the private sector and aims at new cultivars of paprika, jalapeños and hot peppers for processing.

Collection Characterization and Maintenance

During the last five years 654 entries of the germplasm bank were multiplied and characterized at the Embrapa Hortauças, Brasilia, Brazil. Five plants of each entry were grown under greenhouse condition. Five flowers of each genotype were manually auto pollinated, and properly identified at harvest to compare to open pollinated fruits. Seeds were extracted manually, washed in cold water, pre-dried at 32oC for 48h and then dried at 40oC for 48h. After that, seeds were packed in sealed aluminium packs and kept in a cold room at 4oC.

The characterization was done accordingly to the morphological descriptors for Capsicum of the International Plant Genetic Resources Institute (IPGRI). Each entry was characterized for 56 morphological and phenological traits, with some small changes. Plant descriptors: (18 traits): plant height; plant canopy width; leaf width, length, colour, shape, density, pubescence and lamina margin; stem shape, colour, length, diameter and pubescence stem; branching habit; nodal anthocyanin; growth habit; tillering. Inflorescence (13 descriptors): days to flowering; number of flowers per axil; flower position; corolla colour; corolla spot colour; corolla shape; plant male sterility; calyx margin; calyx pigmentation; calyx annular constriction; corolla spot colour; anther colour; filament colour.

Fruit (22 descriptors): days to fruiting; anthocyanin spots or stripes; fruit colour at intermediate and mature stages; fruit position; fruit colour at mature stage; fruit shape; fruit length; fruit width; fruit weight; fruit pedicel length; fruit wall thickness;

fruit shape at pedicel attachment; neck at the base of fruit; fruit shape at blossom end; fruit blossom end appendage; fruit cross-sectional corrugation; placenta length; number of locules; fruit surface; ripe fruit persistence; pungency and aroma. Seeds (3 descriptors): seed colour; number of seeds per fruit; seed surface.

Results and Discussion

Since 1998, 654 of the 1,220 entries of the Capsicum germplasm bank were characterized following the IPGRI descriptors. Of these, 397 were classified as C. annuum, 75 of C. baccatum, 147 of C. chinense and 35 of C. frutescens. The morphological characterization following the above mentioned descriptors allowed the identification of similar species and groups of entries, besides the genetic variability within each species. Great variability was observed for all plant and fruit traits, especially in yield and fruit colour, shape, pungency and aroma. The frequency of 25 selected descriptors was used to characterize 148 genotypes of Capsicum annuum. Corolla spot colour, calyx annular constriction, neck at base of fruit and male sterility were not detected in the evaluated genotypes.

Some IPGRI descriptors were considered as secondary or time-consuming and for these reasons suppressed, while other regarded as important for the breeding program were added, such as pungency, aroma and fruit position. Pungency, aroma and fruit position were included to characterize fruit quality traits more easily. Three persons tasted and smelled small portions of the fruits and made a subjective evaluation (aroma ranging from none to strong; pungency ranging from none to hot). Fruit position was characterized as pendant, erect (upright), intermediate and variable (plants with fruits in more than one position).

All information available (morphological traits, passport data and a digital photo) was grouped into a databank by using the ACCESS software. Some of the entries of the Embrapa Hortaliças Capsicum germplasm bank can be accessed through internet (www.cnph.embrapa.br/projetos/capsicum/consulta/html) at a homepage in Portuguese with a digital photograph, number of registration, origin and species identification. Other relevant information on Capsicum production, history, species and types (with photographs), diseases and pests, and recipes are also available in another homepage in Portuguese (www.cnph.embrapa.br/capsicum).

Eight accessions of local pepper (Capsicum annum L.) landraces were evaluated and characterized under greenhouse conditions in the Jordan Valley at Karama Station during 2001/2002 growing season. Average soil salinity of the experimental field ranged from 6 to 4 dS m-1 at zero to 30 cm and 30 to 60 cm soil depth, respectively. Considerable differences in plant growth parameters, yield and average fruit number plant-1 were observed between the accessions. Accessions with higher shoot and root dry weights such as Jo.209, Jo.202 and Jo.207 and those with higher yields like Jo.204, Jo.209 and Jo.202 are considered less sensitive to stress conditions including salinity. These germplasm can be used by breeders in their breeding programs according to their objectives. Furthermore, The variation in fruit characters: weight, length, width, curvature, pungency and color might be useful for breeders in pepper selection.

Pepper (Capsicum annum L.) is one of the most popular vegetable crops grown in Jordan. Most of pepper cultivars currently grown in Jordan are hybrids. However, some local pepper landraces are still grown in many small farms due to some consumer demand. Local pepper landraces were cultivated under different irrigation water quality including saline water and in different environments for several decades. Therefore, variation between local pepper landraces in response to stress conditions is expected. Wide range of variation was observed in pepper germplasm in several countries including India (Munshi et al., 2000), Albania (Hallidri and Tome, 2000) and Nigeria (Denton et al., 2000). Differences between pepper landraces due to genotype, environments or genotype-environment interaction were also reported (Zewdie and Bosland, 2000). Variations in response to stress conditions and disease and pest tolerance might be of great value for breeders (Zewdie and Zeven, 1997). In Jordan, most of the local pepper landraces are not characterized, Hence this study was initiated to evaluate and characterize some morphological and agronomical traits of eight accessions of local pepper landraces.

Eight accessions of Pepper (Capsicum annum L.) landraces collected from local farmers throughout the country during 1995, and conserved in the gene bank of the National Center of the Agricultural Research and Technology Transfer (NCARTT) were used in this study (Table 1). Plants were grown at Karama Station under greenhouse conditions. Seeds were sown in mid Sep. in 2001 and forty-five day old seedlings were transplanted into greenhouse in late Oct. 2001. Plants were spaced 0.5m apart in 5m raised beds 1.5 m apart. Black polyethylene mulch and a drip irrigation system were used.

The trial was replicated 3 times with 30 plants per replicate and analyzed as a Randomized Complete Block Design (RCBD).

Average plant height, stem color and diameter were recorded at onset of first mature green fruits. Four plants per replicate were harvested; plants were separated into shoot and root and dried at 65oC to constant weight to determine average shoot and root dry weights per plant. Fruits were harvested for several times at commercial ripeness stage starting from Jan.15th 2002 and yield, average fruit number plant-1 and average fruit weight were calculated. The following fruit characters were recorded according to the pepper descriptor published by the International Board for Plant Genetic Resources (IBPGR, 1988), average fruit length and width, fruit curvature, pungency and color.

Variations in vegetative growth parameters were observed between pepper landraces (Table 3). While there were no significant differences in stem diameter among all pepper accessions, plant height ranged from 87 in accession Jo.203 to 146 in accession Jo.209, The heaviest average shoot dry weight (204 gm) was observed in accession Jo.209 with no significant differences from those produced by accessions Jo.202, Jo.205 and Jo.207. The lowest weight, however, was observed in accession Jo.203. Average root dry weight plant-1 ranged from 37 mg in accession Jo.202 to 22.0 mg in accession Jo.109b. Root to shoot dry weight ratio was highest in accession Jo.204 (0.23) while the lowest ratio (0.15) was found in accessions Jo.109a and Jo.109b.

Significant variations in yield were among the landraces (Table 4). Pepper yield ranged from 8.5 ton ha-1 in accession Jo.109b to 27.2 ton ha-1 in accession Jo.204. Average fruit number plant-1 was 30 and 79 in Jo.109a and Jo.204 accessions, respectively. The respective average fruit weight was 8 and 34 gm for accessions Jo.203 and Jo.209. Average fruit length was lowest (6.2 cm) in accession Jo.109b and highest (15 cm) in accession Jo.204 and fruit width was 1.5 and 3.8 cm in accessions Jo.203 and Jo.209, respectively.

Fruit curvature varied from slightly curved to curved. Fruit pungency varied from intermediate to high and fruit color ranged from light green to dark green.

Considerable differences in plant vegetative growth, yield and fruit characters were observed among these accessions (Tables 3 and 4). Though, variation in shoot and root dry weights may be due to differences in genotype of pepper landraces, this does not negate the effect of stress conditions including water and soil salinity. Pepper has been cataloged as moderately sensitive to salt stress; fruit yield decreased by 14% for each unit increase in ECe in the root medium with a threshold value of 1.5 dS m-1 (Maas and Hoffman, 1977). Therefore, pepper accessions with higher shoot and root dry weights such as Jo.209, Jo.202 and Jo.207 and those with higher yields like Jo.204, Jo.209 and Jo.202 are considered less sensitive to salt stress as compared with the other accessions. On the other hand, lower yields in accessions Jo.109a, Jo.109b and the appearance of blossom end-rot in fruits of accession Jo.205 indicate that these accessions are more sensitive to salt stress. The variation in fruit characters: weight, length, width, curvature, pungency and color might be useful for breeders in pepper selection.

As an attempt to find out a way of reducing the labor cost required for harvesting, a breeding program was initiated to develop varieties, which can be harvested once-over. During 1990 to 1992, germplasm collected were evaluated for once-over harvest type (Yoon et al., 1995). In 1993, a cross between desirable resources, 'Jalapeno' and an unknown variety was initiated, followed by subsequent selfing generations in the pedigree breeding scheme from 1994 to 2001. Selection in the field was focused on the possibility of one-time harvesting and good fruit quality.

The first flowering date of 'Saeng-ryeog No. 216' is 87 days from seeding, which is a little later than that of a leading commercial variety, 'Manita'. It has upright fruits, determinate growth habit, and medium fruit size with approximately 10g in fresh weight. On harvesting, the percentage of fully matured and usable fruits was more than 88%, or 437g per plant. The mature fruit color is bright red, and it has no pungency. Field resistance to powdery mildew, anthracnose and virus diseases was medium level in the field experiment conducted in 2001 to 2002.

LOW NIGHT TEMPERATURE EFFECT ON PEPPER OVARIES CHARACTERISTICS. HISTOLOGICAL STUDY

For early season harvest, peppers grown during winter in unheated plastic greenhouses are usually affected by low night temperatures. These conditions have a considerable negative effect on pepper flower development and limit the crop yield.

Under optimal night temperature the flowers are small, whereas under low night temperatures pepper flowers have larger ovaries (Rylski, 1986). Pressman et al. (1998a) reported an increase in ovary size of sweet pepper flowers, as night temperatures decreased from 20 to 12 or 10 °C.

It is suggested that the increased ovary diameter results in the enlargement of cells in the various locations in ovaries developed under low night temperature regime (28/12°C) (Pressman et al.,1998b). Such an alteration of flower part growth is likely to be partly responsible for poor normal fruit set (Tarchoun et al., 2003).

Usually, fruits developed at low night temperature were malformed because of low or no seeds set (Rylski, 1986). Such fruits are without commercial value. Most studies on pepper attribute the lake of fruit set and fruit malformation to the poor pollen viability (Polowick and Sawhney, 1985; Mercado et al., 1997c), investigations on ovules involvement are, however, scarce.

The objective of the present study is to determine histologically the effect of low night temperature in pepper ovaries characteristics especially on the number of ovules and locules.

Plants of hot pepper (C.annuum L.) cultivars Beldi and Baklouti are grown in chambers with day/night temperatures of either low night temperatures regime (LTR: 25°C/10°C) or optimum temperatures regime (OTR: 25°C/20°C).

About 50 flowers from each temperature regime and cultivar were sampled on the day of anthesis, their ovaries were separated and fixed in FAA during 24 hours. After dehydration in different alcohol concentrations and infiltration with paraffin, ovaries were cut transversely with a rotary microtone (Leica) into 10μm sections. These sections were stained with safranine and fast green. For each cultivar and temperature regime, about 20 slides presenting the equatorial sections (Tomer et al., 1998) were used to determine, under a microscope :

1. locules and ovules number per ovary,
2. longitudinal and transversal diameter (Ø1, Ø2) which were parallel and vertical to the ovary circumference, respectively.

To detect the effect of male and female gamete on fruit shape or size, artificial pollination was carried out on Beldi cultivar grow either in low or at optimal night temperature. Pollen, collected from opened flowers developed in the OTR of 20°C, was immediately used to pollinating the flowers (after castration at a bottom stage) of plants grown under the LTR of 10°C (T1-2). A reciprocal pollination was applied. Pollen from lower night temperature was used to pollinating flowers of plants grown under the higher temperature regime (T2-1). About 20 flowers in each replicate (total 60) of each cultivar was treated. Data are means of these replications.

Mature fruits characteristics were recorded for each pollination method: weight (g), length and diameter (mm) and number of seeds per fruit.

Data were analyzed using SAS system, ANOVA was applied at the 5% level.

Ovaries Characteristics

Under the lower night temperature regime (25/10°C) ovaries show a decrease of more than 50% of the number of ovules compared to those developed under the higher night temperature. Therefore, the number of locules records a significant increase ($P<0.05$). Longitudinal and transversal diameters were dependent on the temperature regime. Ovaries developed under the lower night temperature were wider than those under the higher regime, since measurements of transversal and longitudinal diameters (Ø2, Ø1) revealed an increase of 14% and a decrease of 31%, respectively, resulting in the enlargement of ovaries. Pressman et al. (1998b) attributed the swelling of pepper ovaries to the increase of the size of all cells in the ovaries grown under low night temperature of 12°C or treated with triiodobenzoic acid (TIBA).

The effect of LTR on the ovaries characteristics was in a similar magnitude for both cultivars. A marked reduction (about 60%) in the number of ovules was recorded in ovaries of plants grown under 10°C night temperature compared to the control ovaries. Tomer et al. (1998) pointed out that, at night temperature of 12°C, ovules of tomato cultivars were absent from cold affected locules, thus, no seed production could take place. In addition, for the number of locules, differential susceptibility of the cultivars was expressed under LTR. This number was more increased in Baklouti than Beldi ($p<0.05$). The increase of the number of locules was associated to enlargement of ovary particularly in Baklouti. Under low night temperature, transversal diameter (Ø2) of both cultivars was increased, compared to the longitudinal one (Ø1), of about 22% and 16% for Baklouti and Beldi, respectively. Although, at optimal night temperature, the longitudinal diameter was more longer than transversal one resulting in elongated ovaries in both cultivars. Cultivars susceptibility to the effect of low night temperature on ovary abnormalities was noted in others Solanaceae, tomato (Hosoki et al., 1990) and pepper (Pressman et al., 1998b).

Fruit Characteristics

Fruit characteristics, except fruit diameter, were more elevated, in case of self pollination at OTR, than others treatments. Artificial pollination, using pollen from night temperature of 20°C, increase the number of seed per fruit in comparison to the self pollination at night temperature of 10°C and treatment using pollen from lower night temperature (T2-1). Similar results were reported by Polowick and Sawhney (1985) and Pressman et al. (1998a). Although, high significant differences were noted between the number of seeds following self pollination at higher night temperature and the artificial pollination (T1-2) using a pollen from night temperature of 20°C. Theses results indicate that the number of seed seems to be dependant not only to the pollen viability but also to the number of ovules since these ovules were negatively affected by low night temperature of 10°C and thus, artificial pollination with viable pollen (T1-2) was ineffective in improving fruit shape and size.

GENETIC ARCHITECTURE OF CHILLI (CAPSICUM ANNUUM L.)

India is one of the largest producer of chilli (8,32,600 metric tons) occupying third place among the vegetable crops in area and production (Anon., 1997). A genotype with moderate pungency coupled with amenability for cultivation during most of the season is an ideal dual purpose type of commercial exploitation which could be achieved through intervarietal crossing and selection of improved strains in succeeding generations. For which, knowledge on the genetics and mode of inheritance of various economic traits, which influence the yield, is very much essential. Literature on this aspect is limited, hence the present study was undertaken to understand the genetic background in the expression of economic characters through diallel technique. Ten chilli genotypes comprising five pungent and five non-pungent types were crossed in 10 x 10 diallel design in all possible combinations excluding reciprocals during 2001-02. The resultant 45 F1 hybrids were raised along with their parents in randomized block design, replicated thrice. Forty-five days old seedlings were transplanted at 60 x 60 cm spacing keeping 12 plants in each replication for each parents and crosses. Observations on 13 yield and its components were recorded in 5 randomly selected and tagged plants in each of crosses and parents from each replication. Genetic analysis was carried out according to (Hayman, 1954).

The additive component 'D' was significant for all the characters except primary branches per plant. Similarly, the two measures of dominance i.e. 'H1' (dominance effect) and 'H2' (proportion of dominance due to positive and negative effect of genes) were significant for all the thirteen traits. The significant D, H1, and H2 indicated the existence of both additive and dominant gene actions in the expression of these traits. However, the additive component was in higher proportion than dominant factors for plant height, fruit volume, fruit weight, total chlorophyll and total capsaicin content, indicating preponderance of additive gene action. Similar results were also reported by (Ahmed et al, 1982; Rao and Chhonkar, 1983; Sekar, 1984 and Sarala Devi and Arumugam, 1999).

The estimate of 'F' value which indicates the relative proportion of the dominant and recessive alleles were positive for all the traits except fruits per plant and total capsaicin content. This indicated more proportion of dominant alleles in the control of these characters. The environmental factor 'E' was significant only for total chlorophyll which indicated environmental influences in the expression of this trait.

The proportion of (H1/D)1/2 represents the degree of dominance was more than unity for, days to flowering, primary branches per plant, fruits per plant, fruit length, fruit girth, fruit shape index, days to maturity and fresh fruit yield per plant indicating the existence of over dominance for these traits and partial dominance for rest of the traits like, plant height, fruit volume, fruit weight, total chlorophyll and total capsaicin content (Sekar, 1984; Joshi, 1988 and Sarala Devi and Arumugam, 1999).

The ratio H2/4H1 was not equal to 0.25 for all the traits under study indicating unequal distribution of positive and negative alleles. The proportion of dominant to recessive genes in controlling the characters as indicated by the ratio KD/KR was more than one for all the traits except for fruits per plant and total capsaicin content revealing the presence of high proportion of dominant genes in the control of these yield components. The numbers of genes and gene groups controlling the trait (h2/H2) was the maximum for fruits per plant and fresh fruit yield per plant (2-3), while it was less than one for remaining characters.

Moderate to high heritability estimates were observed for all the traits except primary branches per plant. These indicated the larger proportion of additive genes for the inheritance of this traits with less environmental influences. The present study suggested that the recurrent selection followed by pedigree breeding would be helpful to exploit both additive and dominance genes controlling these traits.

DETECTION AND CHARACTERISATION OF SNPS IN CAPSICUM SPP.

In the past few years, SNP (single nucleotide polymorphism) discovery and marker development have increased exponentially across a wide range of plant species. SNP markers offer great potential for wide genome coverage and high throughput analysis in breeding through marker-assisted selection (MAS) and other studies.

The identification of SNPs for targeting quality traits (QTLs) as well as disease resistance genes is already advanced in a number of major crops. We report on a preliminary study aimed at the detection of SNPs, by PCR-SSCP (single strand conformation polymorphism), in genes coding for well-studied enzymes in Capsicum spp. In a first screening, PCR products were analysed by SSCP for the presence of point mutations in amplicons from different samples. Polymorphic fragments were then sequenced for SNPs and validated by tetra-primer ARMS (amplification refractory mutation system) PCR (Ye et al., 2001).

Genomic DNA was extracted (Doyle and Doyle, 1987) from seven Capsicum species (C. annuum L., C. chinense L., C. frutescens L., C. pubescens L., C. chacoense L., C. baccatum L. and C. tovari L.), together with 17 C. annuum L. accessions. All the materials were obtained from the germplasm bank of Di.Va.P.R.A, Plant Genetics and Breeding. Fifteen genes from C. annuum were selected from the NCBI database for SNP mining. A set of primers were designed for each gene to amplify fragments of about 450 bp. Each fragment was chosen from a region overlapping the intron-exon junctions.

PCR reactions were performed at 5 min at 94°C, 35 cycles of 30 s at 94°C, 30 s at 55°C, 1 min at 72°C followed by 5 min at 72 °C. SSCP analysis (amplification and electrophoresis) was performed as described by Sunnucks et al. (2000) with slight modifications. Gels were silver stained using the procedure described by Bassam et al. (1991).

The polymorphic fragments were purified and directly sequenced in an ABI 3100 sequencer using BigDye™ terminator chemistry. The sequences were aligned with Megalign (DNASTAR) using Clustal algorithm. Putative SNPs were identified. Genetic variability in DNA sequences was measured by the nucleotide diversity, p, with p = K/L (K=number of differences per nucleotide site and L=sequence length on bp (Nei and Li, 1979)

Thirteen of fifteen fragments were amplified as unique scorable bands. Eight of fifteen fragments were analysed by SSCP. Data obtained provided a preliminary estimation of the polymorphism present in both Capsicum species and C. annuum accessions. Sequencing, performed in samples shown to be polymorphic to SSCP analysis, gave evidence of putative SNPs. A total of 2719 bp was scanned, of which 1836 bp were derived from coding region (exons) and 883 bp from non coding region (promoters and introns).

Fifty-seven SNPs were found, of which 34 were in coding region (cSNPs), however only 18 of them caused amino acid changes. Most of the mutations were at the third position of the codon (47%) while 38% and 18% were at the second and the first position respectively.

The frequency of base substitutions, between 5 of the 7 Capsicum species in study, was 1 SNP per 48 bp while 1 SNP per 1360 bp between C. annuum accessions. In previous studies it was observed a frequency of 1 SNP per 1 x 103 bp in mouse (Lindblad-Toh et al., 2000) and 5 x 104 bp in man (Wang et al.,1998). Our results show that, among the detected SNPs, 50% were transitions and 50% transversions.

This is not in full agreement with data obtained in Beta vulgaris (Schneider et al., 2001), mouse and man (Wang et al.,1998; Lindblad-Toh et al., 2000) but might be explained by the relatively low number of scanned nucleotides. Nucleotide diversity (p) ranged between 3.8 and 45.5 (average 19.2) among Capsicum species and between 0 and 1.9 (average 0.7) among C. annuum accessions.

Sequencing data were validated by tetra-primer ARMS-PCR procedure. Among the three SNPs detected in C. annuum, one SNP was confirmed at base 512 of the capsanthin/capsorubin synthase gene and two in EG2 gene at 5182 and 5252 base position respectively. The SNP EG2-5252 showed to be heterozygous and it made possible to distinguish the C. annuum L. Thai accession from all the others.

The two step strategy used for SNP mining using SSCP and sequencing followed by the tetra-primer ARMS-PCR genotyping proved to be a reliable and cost effective method for SNP characterisation. The use of SSCP offers some economic advantages as it is possible to identify and sequence only the polymorphic amplicons. Nevertheless a complete study of nucleotide variability in this species would require a large genonome coverage.

Conclusions and Perspectives

This on-going research was performed to obtain a preliminary view of pepper nucleotide diversity. Since SNPs appeared to be present with a slightly higher frequency in non coding regions than in coding ones, future research will focus on a deeper analysis of this findings. SNP mining will provide invaluable tools for application in MAS, genetic diversity studies, germplasm characterisation and variety identification.

NOTE ON EVALUATION OF CHILLI (CAPSICUM ANNUUM L.) GENOTYPES FOR BIOCHEMICAL CONSTITUENTS

Chilli (Capsicum annuum L.) is the only source of capsaicin-an alkaloid that is a digestive stimulant, an important ingredient of daily diet and a cure for rheumatic troubles. Chilli also contains many other medicinal properties. Green chilli fruits are valuable on account of their richness in ascorbic acid, which is an important vitamin. The fruit colour is due to the presence of total carotenoids pigments, which mainly consists of capsanthin and capsorubin. The extent of colouring matter is important for spice industry. Furthermore the colouring matter is used to impart colour to the other food products. Keeping view of the above mentioned points, thirty chilli genotypes were evaluated in the present study for their biochemical constituents viz. capsaicin, ascorbic acid and total carotenoids.

The ascorbic acid content was determined by 2, 6 dichlorophenol indophenol method of AOAC (1970). Estimation of total carotenoids was done as per the method given by Ranganna (1997). Capsaicin content was found out as per the method given by Quagliotti (1971). Fruit width was estimated by using Vernier calipers.

The data regarding capsaicin, ascorbic acid and total carotenoids contents are given in the table 1.Capsaicin content showed a narrow range of variation from 0.33mg/100mg in Rajasthan local to 0.49mg/100mg in KDCS-810 with an overall mean of 0.41mg/100mg.Smaller sized fruits which are short and thin contained more capsaicin compared to long fruits. The mean ascorbic acid content was lowest in DCL 344 (78.30 mg/100g) and highest in ACS 2000-02 (188.0mg/100g). Ascorbic acid content had a grand mean of 130.01mg/100g of fresh fruit weight. Long fruits contained higher ascorbic acid content. Total carotenoids showed a range or variation from 1475.3μg/

100g in DCL 008 to 4208.0μg/100g in Red star with an over all mean of 2732.4μg/100g.

From the results it may be concluded that significant variation exists among the selected chilli genotypes for the estimated biochemical constituents. The smaller sized fruits were associated with higher capsaicin content and large fruited genotypes contained higher ascorbic acid. Similar association was also reported by Sharma et al (1981).

VARIABILITY, CORRELATION AND PATH ANALYSIS IN KHARIF GROWN CHILLI (CAPSICUM ANNUUM L.) GENOTYPES FOR DIFFERENT CHARACTERS

Estimates of variability, heritability, genetic advance, correlation and path co-efficient analysis were carried out in 26 chilli genotypes in transitional tract of Karnataka. High degree of genotypic and phenotypic co-efficient of variation was observed for number of primary branches, fruit length, pericarp thickness, number of fruits per plant and green fruit yield per plant. High heritability coupled with high genetic advance as a percentage of mean was observed in respect of fruit length and green fruit yield per plant. Yield per plant showed positive correlation with number of fruits per plant and pedicel length. Path co-efficient analysis revealed the importance of number of fruits per plant in determining selection criteria for improvement of chilli yields.

Chilli is grown in arid and semi-arid regions of India. Therefore it is important to evolve varieties for maximize economic yield and consistent in their performance in a given environment. The gain from selection in crop breeding program is dependent on the amount of variability for economic characters in the germplasm. For rapid progress in selection of the germplasm study on association of yield with component characters shall help in elucidating the intrinsic nature of correlation and ease the selection procedures.

The field experiment was conducted during kharif 1999 at the Department of Horticulture University of Agricultural Sciences, Dharwad with 26 genotypes of chilli using Randomized

Block Design in three replications. A spacing of 60 x 30 cm. was followed. Observations were recorded on five randomly selected plants for 13 characters viz., plant height (cm), number of primary branches, plant spread (cm), days to 50 per cent flowering, fruit length (cm), pedicel length (cm), pericarp thickness (cm), number of seeds per fruit, green fruit weight (g), number of fruits per plant, green fruit yield per plant (g) and fruit rot incidence (%).

The mean values of five plants were used for analyzing genetic coefficient of variation, heritability (in broad sense), and genetic advance as per cent of mean were computed by using standard methods along with correlation and path analysis as per method followed by Dewey and Lu (1959).

It is evident from the data there was wide range of variability for all the characters providing an ample scope for selecting the desirable types.

The characters viz., number of primary branches, fruit length, pericarp thickness, number fruits per plant and green fruit yield per plant showed highest degree of PCV and GCV indicating broad genetic base for these traits (Rani and Singh, 1996). High variability for fruit rot disease incidence indicate high scope for selection against the disease. The highest estimates of heritability was obtained for plant height (93.7 %), number of primary branches (91.7%), fruit length (95.7%) and green fruit yield (90.5 %) along with high genetic advance over mean. Thus, improvement through selection could be made for these characters while,

days to 50 per cent flowering and plants spread showed high heritability coupled with low genetic advance over mean. This indicates predominance of non-additive gene effects which arises when variability of the traits is low (Kataria etal., 1997).

Correlation studies revealed that green fruit yield had positive and significant association with number of fruits (Jose and Khader, 2002) and pedicel length. Green fruit weight was significantly correlated with number of fruits per plant in negative direction. On the contrary number of fruits per plant seem to be major character contributing to the yield per plant. Hence, care to be taken to strike balance while selecting for the characters based on market demand and preference.

Path analysis provides information on real nature of association by separating into direct and indirect effects. Among the character studied number of fruits had maximum direct effect (0.671) on yield followed by pedicel length and pericarp thickness. Low negative direct effect (-0.073) of fruit weight was observed with yield per plant due to indirect negative association with number of fruits per plant. However, overall correlation remained positive due to the influence of other characters studied (Sharma and Roy, 1995). Hence from the investigation it could be concluded that number of fruits per plant and pedicel length deserve greater weightage during selection for yield in chilli varieties.

VARIABILITY IN CAPSICUM CHINENSE JACQ. GERMPLASM ELIZABETH V. CHERIAN AND P. INDIRA

Capsicum chinense Jacq is an under exploited species of the genus Capsicum. The perennial species is ideal for the homestead cultivation in Kerala, because of its ability to yield substantially under shaded conditions and tolerance to pest and diseases. It is locally known as Malimulagu as it is being exported to Maldives. As there is much diversity for this species in Kerala, there is good scope for selection. The present study was taken up to assess the genetic variability existing in this crop.

Twenty five diverse accessions of Capsicum chinense were used for the study during two seasons. All of them were local collections. The experiment was laid out in randomized block design with two replications. The crops were raised during October 1997 to June 1998 and September 1998 to April 1999, following the recommendations of Package of Practices of Kerala Agricultural University 1996.

Observations were recorded on biometric characters like plant height, number of primary branches, days to first flowering, days to harvestable maturity, pedicel length, fruit length, fruit girth, fruit weight, number of seeds per fruit, number of fruits per plant, yield per plant and driage percentage. Quality attributes like capsaicin content, colour value and oleoresin were also recorded. Bacterial wilt incidence was also noted.

Results and Discussions

The results of analysis of variance of the accessions of Capsicum chinense showed significant difference for all the 16 characters studied during both seasons.

Plant Height

The height ranged from 29cm to 52cm during the second season. CC26 was the tallest during first season while CC20 was the tallest during second season. CC38 was the dwarfest during both seasons.

Number of Primary Branches

Number of primary branches ranged from 2.8 (CC26) to 10.8 (CC37) during first season while in the second season it ranged from 2.7(CC 26) to 11.8 (CC37).

Days to First Flowering

CC28 was the earliest to flower (104.5 days) during first season while CC30 was the latest (121.5). During second season CC16 and CC23 were the early flowering accessions and CC17 was late flowering.

Days to Harvestable Maturity

CC23 was the earliest to attain harvestable maturity (147days) during the first season. While CC10 (138days) was the earliest during second season. CC17 was the latest during both seasons.

Pedicel Length

Pedicel length varied from 1.35cm (CC38) to 4.25cm (CC21) during the first season and from 1.40 cm (CC38)to 4.35cm (CC21, CC15) during the second season.

Fruit Length

Fruit length varied from 0.9cm to 6.2 cm in the first season and from 0.95cm to 6.15 cm in the second season. CC17 was the lung fruited genotype during both seasons while CC20 was the short fruited.

Fruit Girth

Fruit girth ranged from 3.05cm (CC38) to 8.5cm (CC2) in the first season and from 3.00cm (CC37, CC38) to 3.0 cm (CC2) during the second season.

Number of Seeds per Fruit

Number of seeds ranged from 20.5(CC38) to 36.5(CC23) during the first season and from 28.9(CC38) to 53(CC2) during the second season.

Fruit Weight

Average fruit weight ranged from 0.9 to 6.6 g and 0.9 to 7.2 g respectively during the first and second season. Fruit weight was highest in CC17 and slowest in CC20.

Number of Fruits per Plant

The number of fruits per plant ranged from 4.0(CC2) to 52.5 (CC37) during first season and from 5.0(CC9) to 63.5(CC8) during second season.

Yield per Plant

During first season yield per plant ranged from 12.0g (CC38) to 75.5(CC8). The yield varied from 14.0g (CC38) to 185.0g (CC8) during second season.

Driage

The driage varied from 20.75 to 24.85 during first season and from 20.3 to 24.6 during second season.

Bacterial Wilt Incidence

Bacterial wilt caused by Ralstonia solanacearum ranged

from zero (CC32) to 46 percent (CC1,CC3, CC17) and from zero (CC28) to fifty percent (CC16) in the second season.

Oleoresin

The oleoresin content ranged from 8.55 percent to 25.65 percent during first season and from 8.80 percent to 24.05 percent during second season CC17 had the maximum oleoresin content.

Colour Value

Nesslerimetric colour value ranged from 579.50 (CC25) to 1433.5 (CC37) in the first season and from 549.0 (CC25) to 1533.5 (CC51) in the second season.

Capsaicin

Capsaicin content of the accession ranged from 0.80 per cent (CC20) to 1.85 per cent (CC2, CC5) during first season and from 0.80 per cent (CC20) to 1.75 percent (CC5) during second season.

The existence of considerable variation in Capsicum chinense germplasm indicated the scope for improving the scope for improving the population for these characters wide range of variability in Capsicum annum for biometric characters were reported by many workers (Arya, 1979, Nair et al, 1984, Singh et al,1998).

VARIABILITY, HERITABILITY AND GENETIC ADVANCE IN BIRD PEPPER (CAPSICUM FRUTESCENS L.) I. SREELATHAKUMARY AND L. RAJAMONY

Bird pepper (Capsicum frutescens L.) is an economically important species, valued for its perennial nature and highly pungent small fruits with characteristic flavour. It has high capsaicin content and ideal for extraction of high pungency oleoresin. Almost all the bird pepper types are indigenous and characterized by a wide range of observable variability. Breeding works in Capsicum have so far been concentrated mainly on C. annuum with little emphasis on C. frutescens. Considering the importance of bird pepper in industry and export an experiment has been taken up to estimate the genetic variability, heritability and genetic advance in the available accessions of bird pepper.

This experiment was carried out at the College of Agriculture, Kerala Agricultural University, Thiruvananthapuram, Kerala, India, during the period from 1998 -1999. The materials for the study consisted of 20 accessions of bird pepper collected from different parts of Kerala. The experiment was conducted in randomised block design with two replications. Ten plants were maintained per plot. All the recommended cultivation practices were followed to raise the crop under irrigated condition. Five plants were selected randomly from each accession and observations were recorded on plant height, internodal length, stem girth, leaf area, petiole length, days to first flower, node to first flower, height of node to first flower, fruits per plant, fruit length, fruit girth, fruit weight, yield per plant and incidence of mite. The analysis of variance (Panse and Sukhatme, 1978), variability for different quantitative characters (Burton, 1952) and expected genetic advance at 5.0 per cent intensity of selection (Johnson et al., 1955) were calculated

Analysis of variance showed significant differences among the accessions for plant height, stem girth, leaf area, leaf petiole length, fruits per plant, fruit length, fruit girth, fruit weight and yield per plant.

The present investigation revealed considerable amount of variation for all the characters studied. Higher values of genotypic (3250.51) and phenotypic (3333.35) variances were recorded for yield per plant.

GENE ACTION AND COMBINING ABILITY FOR FRUIT YIELD AND ITS COMPONENT CHARACTERS IN SWEET PEPPER

Sweet pepper (Capsicum annuum L.) is an important vegetable cultivated in most parts of the world for its green, immature fruits used in various preparations and salads. The crop grows well in cool weather, which makes Kashmir valley the most ideal place for cultivation of not only fresh green fruits but the seeds of varieties and hybrids could be produced during summer on large scale for export. Despite such potential only limited cultivars are under cultivation, which are either low yielders or poor in quality.

This necessiated search for high yielding varieties or hybrids with superior quality. Prior to initiation of any improvement programme, the knowledge of combining ability of parents and crosses and the gene effects involved in the inheritance of various traits is a prerequisite as it provides a guideline for selecting elite parents or combiners which may later be hybridized either to exploit heterosis or to accumulate fixable genes through selection and also helps in understanding inheritance of various quantitative characters which are essential for choosing appropriate breeding methods. The current investigation is an attempt to get information on nature and magnitude of gene action and combining ability by using line x tester analysis designed by Kempthorne (1957).

The experimental material comprised eight diverse lines of sweet pepper (KSPS-464, HC-202, KSPS-4, World Beater (WB) KSPS-461, KSPS-13, Vinedale and HC-201) each crossed to three testers namely California Wonder (CW), Oskash and KSPS-2 during 1996 tc develop 24 F1s. The 11 parents (8 lines and 3 testers) along with 24 F1s were evaluated during 1997 at Vegetable Experimental Farm, SKUAST, Srinagar in a randomized block design with three replications. In each replication 10 plants each of F1s, and parents were planted in a single row at a spacing of 60 cm between rows and 45 cm between plants within the row. The observations were recorded on five randomly selected plants of each entry for ten different characters. The combining ability was carried out as per the method given by Kempthorn (1957). The additive (s2A) and non additive (s2D) genetic variances were estimated from the mean squares for general

combining ability (gca), specific combining ability (sca) and error. Average degree of dominance (ADD) was worked out as(s2D)/(s2A)½ given by Comstock and Robinson (1952)

The significant gca and sca variances and the estimated genetic components of variances (Table 1) revealed the importance of both additive (s2A) and non-additive (s2D) genetic variances in the inheritance of various characters. However the additive component of variance (s2A) was more prominent in the genetic control of days to first fruit set which was further supported by degree of dominance whose value was less than one (0.5015). Whereas non additive gene effects (s2D) played a greater role in the inheritance of plant height, plant spread, branch number, fruit girth, fruit number, fruit weight and fruit yield. Importance of non-additive component of variance has been reported by Sharma and Saini (1977) and Ahmad et al (1994). The average degree of dominance being more than one for these characters was also in agreement with the above results which indicated over dominance coupled with non additive gene effects and such genes could be successfully exploited by employing heterosis breeding. For fruit length and pericarp thickness both additive and non additive gene effects were almost of equal magnitude and their degree of dominance value being 0.92 and 0.99 respectively, indicated partial to complete dominance which could be exploited by following recurrent selection as it is most ideal for utilizing both type of gene actions.

For different characters, different parents proved as good combiners. Among parents HC-202 and Oskash for days to first fruit set. KSPS-461, KSPS-464 and Vinedale for plant height; Vinedale, KSPS-4 and KSPS-464 for plant spread; KSPS-461, KSPS-4 and KSPS-13 for number of branches per plant; KSPS-461, KSPS-464, and W.B. for fruit length, KSPS-13, W.B. and C.W. for fruit girth KSPS-13, KSPS-464 and KSPS-2 for pericarp thickness; KSPS-461, Vinedale and Oskash for number of fruits per plant; KSPS-461, KSPS-13 and KSPS-2 for weight per fruit and KSPS-461, KSPS-13 and KSPS-2 for fruit yield exhibited desirable significant gca effects and proved to be good general combiners.

An overall view of the gca effects indicated that the parents viz KSPS-461, KSPS-13 and KSPS-2 in general proved as good general combiners for most of the traits and could be used in hybridization as they possessed high gca effects for fruit yield and yield attributing characters such as weight per fruit, fruit number, fruit length and branch number. It was also evident that for most of the characters the parents with good general combining ability possessed high mean values but some parents like KSPS-2 and KSPS-4 though highest yielders but were poor general combiners for yield. This revealed that combining ability cannot be always judged accurately only by per.se performance especially like fruit yield which is dependent on many metric traits and has a polygenic control. For selection of parents one should therefore consider both gca effects as well as per.se performance for achieving desired results.

Specific combining ability effects presented in Table-3 revealed that none of the hybrids exhibited desirable significant sca effects for all the traits. Out of 24 cross combinations, 10 crosses had significant desirable sca effects for plant height, 9 for plant spread, 6 for number of branches per plant, 7 for fruit length, 9 for fruit girth, 5 for pericarp thickness 9 for fruit number, 12 for weight per fruit and 8 for fruit yield. Besides revealing significant desirable sca effects most of these crosses also exhibited more or less high per se performance.

Out of 24 hybrids, the best specific combinations were KSPS-461 x KSPS-2 and Vinedale x C.W for plant height, KSPS-464 x Oskash and HC-201 x KSPS-2 for ps; HC-201 x KSPS-2 and Vinedale x Oskash for number of branches; KSPS-13 x CW and HC-201 x KSPS-2 for fruit length; HC-201 x Oskash and W.B x KSPS-2 for fruit girth; W.B x KSPS-2 and KSPS-4 x Oskash for pericarp thickness; Vinedale x CW, KSPS-13 x C.W and W.B x Oskash for fruit-number; HC-201 x KSPS-2, KSPS-461 x Oskash and W.B x KSPS-2 for weight per fruit and HC-201 x KSPS-2, vinedela x C.W. and KSPS-461 x Oskash for fruit yield. Besides revealing significant desirable sca effects, most of these crosses also exhibited more or less high per se performance which is result of either poor x poor or poor x high general combining parents.

In the highest yielding crosses KSPS-461 x Oskash, Vinedale x C.W. and HC-202 x Oskash, the parents Oskash, HC-202 and Vinedale were the poor combiners while the parents viz KSPS-

461 and C.W. had good general combining ability. Since most of the high yielding crosses had one parent with high gca effect, it seems both additive and non-additive genetic components are playing an important role in the inheritance of fruit yield. This is in agreement with the observations made by Gopalakrishnan et al (1987) and Ahmed et al (1997) who indicated suitability of such cross combinations for extracting transgressive segregants. It was also evident from the results that the crosses namely KSPS-461 x Oskash, HC-201 x CW; Vinedale x CW; HC-202 x Oskash and KSPS-461 x KSPS-2 besides having high sca effects also recorded superior per se performance inrespect of fruit yield and most of the yield attributing traits and thus could be considered for the commercial exploitation of hybrid vigour under temperate region of our country.

ESTIMATION OF HETEROSIS, COMBINING ABILITY AND PER SE PERFORMANCE IN SUMMER GROWN CHILLI (CAPSICUM ANNUUM L.) FOR YIELD AND RESISTANCE TO LEAF CURL COMPLEX

The productivity of chilli is hampered by poor quality and susceptibility to leaf curl complex. Among the three seasons during which the crop is cultivated, summer yields are low on account of above said problems.

Substantial solution to the constraints is the necessity to have a security of the crop and quality of the produce through heterosis breeding. Therefore, the recognition of the genetic attribute available in some of the genotypes used in the experiment and evaluation of parents and F1's was carried out during summer season. This would help in selection of superior parents and sca effects for superior hybrids.

Six diverse chilli genotypes were crossed in diallel mating fashion to obtain 15 hybrids. The hybrids were evaluated along with their parents and commercial check in a randomized block design with three replications at Department of Horticulture, UAS, Dharwad during summer 1999 with a spacing of 60 X 45 cm. Observations were recorded on days to 50% flowering, fruit length (cm), fruit diameter(cm), pedicel length (cm), pericarp thickness(cm), number of seeds per plant, green fruit weight (g), number of fruits per plant, green fruit yield per plant(g) and leaf curl complex (%).

Statistical analysis of combining ability based on mean values was done as per method I and method II of Griffing (1956). The F1 hybrid performance was calculated as the estimate of heterosis over better parent. Pooled gca status of the parents was analyzed by compiling gca effects of different characters studied as per the method suggested by Arunachalam and Bandopadhay (1979). Top crosses were considered based on per se values.

The perusal of analysis of variances revealed significant differences among all the characters thus, indicating wider genetic variability among the genotypes in the present investigation. GCA and SCA variances for ten characters were studied. The GCA and SCA variances were significant for all the characters except pedicel length and pericarp thickness respectively.

The magnitude of estimated components of dominant variance was more than additive variance for days to 50% flowering, fruit diameter, pedicel length, green fruit weight, number of fruits per plant, green fruit yield and resistance to leaf curl complex indicating predominance of additive gene effects. This is in conformity with the findings of Basavaraj (1997) and Mulge (1992). Therefore, heterosis may be rewarding for the aforesaid characters. Since the estimates of additive gene effects were higher in fruit length and number of seeds per fruit.Progeny selection would suffice in accumulation of desirable genes for these characters. Additive and dominance variances were of equal magnitude for pericarp thickness, in such situation reciprocal, recurrent selection may be adopted for population improvement.

From the study on gca effects none of the parents were good general combiners for all the characters however, KTPI-19 (6.20, Pant C-1 (14.41) and RHRC-Cluster-Erect (-8.80) exhibited highest gca effect for green fruit weight, green fruit yield per plant and resistance to leaf curl complex, respectively.

In the present investigation, fruit weight, fruit yield and resistance to leaf curl complex were given more emphases as a measure of productivity in summer grown chilli. Per se performance, heterosis and sca effects were considered together. In case of fruit weight the heterosis and sca effects were negative though per se values of the top crosses were high. The negativity or insignificance was due to wide variability involved in the parents of the crosses for the trait viz., RHRC-Cluster-Erect X KTPI-19 and Pant C-1 X KTPI-19 (Lohithaswa et al.,2001). Similarly, insignificant heterosis was observed when LCA 206 (105.53g /plant) was crossed with KTPI-19 (182.3 g/plant) for green fruit yield per plant. Hence, the chances of F1 heterosis enhanced when parents were selected from intermediate classes of per se values. This was evident from the mean values of the parents of top two crosses for yield per plant as they showed significant positive heterosis and sca effects.

With regard to resistance to leaf curl complex, the heterosis values when considered alone were found misleading but there was a correspondence between sca effect and per se values in required (negative) direction. Therefore, selection for yield and leaf curl complex in chilli could be simplified by selecting the parents based on mean values.

The study revealed that the frequency of heterotic hybrids were comparatively high when one of the parents(female) involved in the crosses was of low combining ability status. This could be due to complementation of differing alleles or genes and implies that mostly non additive variation plays a major role in expression of green fruit weight, green fruit yield per plant and leaf curl complex resistance.

QUANTITATIVE CHARACTERS IN RED PEPPER CULTIVARS FOR GRINDING (CAPSICUM ANNUM L.)

Many of the characters describing pepper are in particular correlation. The quantitative characters such as number of fruits per plant, length, diameter and average weight of the fruit have a great economic importance but their evaluation and analysis impede the work of the breeder. The evidence of strong correlation between them made easy the breeding process.

Many researchers look for correlation between different characters in pepper (Depestre et al., 1989; Ahmed et al., 1997; Jose & Khader, 2002; Pinaki et al., 2002). The objective of this study is to establish what is the correlation between the morphological characters in the main cultivars of red pepper for grinding.

The study is carried out in the experimental field of the Institute of Horticulture and Canned Foods, Plovdiv (former the "Maritsa" Vegetable Crops Research Institute) during the period 1996-1999. Six cultivars of red pepper for grinding: the Bulgarian ones 'Gorogled 6' and 'Buketen 50', the Spanish 'Belrubi' and 'Negral' and the Hungarian - 'Kaloscai 801' and 'Mihalytekeli' were tested.

The trial is performed in alluvial-meadow soil by block method in four replications and experimental plot of 4,8 m2 according to the adopted technology for cultivation of red pepper for grinding (Veselinov et al., 1984). Twenty plants and fruits from each treatment have been analyzed. The following morphological characters: plant height (1), stem height (2), fruit number on the main stem (3), fruit number on the branches (4), fruit length (5), fruit diameter (6), pericarp thickness (7), locules (8), average fruit weight (9) and usable fruit part (10) have been observed. The cultivars are sweet peppers (paprika) intended for grinding and distinguish by growth type, habit, orientation and shape of the fruits.

The correlation coefficients for the studied period of morphologic characters separately for each cultivar in most cases are insignificant during the separate years of investigation. Probably this is due to very little variability in the studied characters in each cultivar. Only the correlation between the average fruit weight and the weight of the usable part in all cultivars and all years of cultivation is strong. The total correlation for these two characters logically is very strong, r = 0.951. In this case could be considered that the link is universal. The presence of such correlation of tested cultivars, which are comparatively uniform in morphology, is a proof that small changes in one character cause adequate alteration in the other. This correlation has also very important practical aspect because it gives an opportunity with comparatively good accuracy by the weight of the harvested produce to be predicted the weight of the usable part which will enter for drying and further processing.

CORRELATION AND PATH COEFFICIENT ANALYSIS IN BIRD PEPPER (CAPSICUM FRUTESCENS L.)

Yield in plants is the end product of interaction of many correlated characters. Selection for these characters will be more effective when it is based on component characters that are highly heritable and positively correlated. When more number of variables are considered in correlation, the association becomes more complex and less obvious. The use of path coefficient analysis is helpful under such situation. This analysis shows the direct and indirect associations and reveals the most reliable yield contributing characters. In bird pepper (Capsicum frutescens L.), a wide range of variability is available which provides a great scope for improving fruit yield through a systematic and planned selection programme for one or more direct or indirect yield components. Keeping in view the above facts, the present investigation was conducted to determine the nature and degree of association among the characters and their direct and indirect effects on yield.

Materials and Methods

Twenty accessions of bird pepper (Capsicum frutescens) collected from different parts of Kerala were evaluated at the College of Agriculture, Thiruvananthapuram, Kerala, India during 1998 - '99. The experiment was conducted in randomised block design with two replications. Ten plants were maintained per plot. Five plants were selected randomly from each accession and observations were recorded on plant height, stem girth, leaf area, days to first flower, fruits per plant, fruit length, fruit girth, fruit weight and yield per plant. Correlation of various biometrical characters was undertaken as per the procedure suggested by Singh and Choudhary (1979). Path coefficient analysis using genotypic correlation coefficient was carried out according to Dewey and Lu (1959).

Result and Discussion

In all the characters, genotypic correlation coefficient was found to be higher in magnitude than phenotypic correlation coefficient indicating a strong inherent association among various characters (Table 1). Similar observations were made by Sundaram and Ranganathan (1978), Rao and Chonkar (1981) and Choudhary et al. (1985).

The phenotypic and genotypic correlation coefficient revealed that the association of plant height with other traits was low. The positive association of leaf area with days to first flower, fruit length, fruit girth, fruit weight and yield per plant suggested that the selection of genotypes with high leaf area would help in isolating lines with early flowering and high yield. Days to first flower were found to be negatively correlated with fruits per plant.

The results are in conformity with the findings of Ahmed et al. (1997). Yield per plant showed highly significant positive correlation with fruits per plant, fruit length, fruit girth and fruit weight suggesting that these characters are the most important yield components and that effective improvement in yield can be achieved through selection based on these characters. Fruit length and fruit girth showed significant positive correlation with fruit weight. Sheela (1998) reported significant positive association of economic traits like number of harvests, fruit girth, fruit length, mean fruit weight and fruit size with yield in C. frutescens.

The results of path analysis revealed that fruits per plant, fruit length, plant height and stem girth had shown positive direct effect on yield while leaf area, days to first flower, fruit girth and fruit weight had exerted negative direct effect on yield. The direct effect of fruits per plant was positive and much higher than its genotypic correlation with yield. It exerted positive indirect effect through fruit length while its contribution through fruit girth and fruit weight was negative. Positive direct effect of number of fruits was supported by Gill et al., (1977), Sundaram and Ranganathan (1978) and Munshi et al., (2000) in chilli. The direct effect of fruit length was also positive.

Its indirect effect through fruits per plant was high and positive indicating that direct selection for fruit length and indirect selection for fruits per plant can increase yield. Though the direct effect of fruit girth and fruit weight was negative, its indirect effect through fruits per plant and fruit length was high and positive indicating that indirect selection for fruit length and fruits per plant can increase yield.

The direct effect of plant height was positive and much higher than its genotypic correlation with yield. It exerted high and positive indirect effect through fruits per plant and fruit length. The low residual effect (0.1057) indicated that all the important characters are correlated with yield. Rao and Chonkar (1981) and Munshi et al., (2000) also observed low residual value on their study. Based on correlation and path analysis studies, it could be concluded that selection for fruits per plant, fruit weight, fruit length and fruit girth might lead to increase in yield.

10

Pollen Cryopreservation in Capsicum Species

INTRODUCTION

Hot pepper is one of the most important commercially grown vegetables in the tropics, and is probably most important after tomato (Grubben, 1977). The fruits are popular for its nutritional value and used in cuisines all over the world. One of the important aspects concerning crop improvement of capsicum is disease resistance. Many of the domesticated capsicum varieties are susceptible to an array of diseases such as bacterial wilt, root knot, viruses and bacterial leaf spot. The strategy of a capsicum breeder is to assemble into cultivars with genetic potential for yield, disease resistance and improved quality. Modern pepper breeding has relied on a relatively narrow gene base within various cultivars groups, despite the morphological genetic diversity apparent both intraspecifically and interspecifically.

Under proper storage conditions pepper pollen can maintain viability for extended periods. Sources of resistance are available for creating diverse genepools. Hitherto no reports are available on pollen cryopreservation in Capsicum species. Rylski (1986) reported that in Capsicum, seed set and production could be improved by pollinating with large quantities of pollen. A consolidated stock of viable pollen belonging to different species/ cultivars kept conserved in a pollen cryobank will help breeders in achieving this objective. This study reports the feasibility of cryopreserving pollen belonging to 4 capsicum species namely

C. annuum L., C. baccatum L. C. chinense Jacq. and C. praetermissum Heiser & Smith.

Materials and Methods

Pollen Collection: Flowers were collected between 10 AM and 11 AM from field grown healthy plants and brought to the laboratory. Pollen was extracted by keeping the flowers in a BOD incubator attached with fluorescent light at 250C for dehiscence, scooping out the pollen mass through lateral sutures of the anther with a spear needle. Pollen thus extracted is collected in a clean butter paper.

Viability& Sterility Assessment: Methods followed are described in detail (Alexander et al 1991). Pollen was germinated by improved cellophane method. The preparations were incubated at 25±20C for a durations of 4-6 hours, after which staining was accomplished using the versatile stain (Alexander, 1980), pollen stained green are recorded sterile.

Cryopreservation: Pollen samples assorted specie-wise were bulked and transferred into empty gelatin capsules and packed in a laminated aluminum pouch and sealed air tight. The pouches were put in 'Eppendoff' tubes, capped and sealed with adhesive tape and held in liquid nitrogen vapours in a Cryogenic container (MVE- Mach SM-33) with the help of a Teflon string. After 48 hours of cryogenic exposure, viability of pollen samples was re-assessed.

The data was analyzed by Factorial CRD with arcsine conversions of percentages for germination values and ANOVA was performed.

Results and Discussion

The initial germination profiles recorded for the 4 capsicum species in this study was below 55 per cent (as per arcsine conversions). Pollen samples responded well to the 48-hour cryogenic exposure. There was no significant difference in germination percentages of fresh and stored pollen (presented in table-1) except in C.chinense ($p=0.05$) where a reduction in post-cryogenic exposure viability was observed.

C. praetermissum pollen although low in fresh pollen viability, recorded a significant increase in germination profile.

Sterility was minimum in C. baccatum and C .chinense, having a relatively higher fresh pollen germination profile. Conversely, highest pollen sterility was recorded for C. annuum. Wit low germination value. With increasing trend in sterility, lower fresh viability profiles were observed.

The results described above have special significance to capsicum breeders and Gene bank curators. Breeders can avoid growing a crop exclusive by for pollen collection, once sufficient NGD is kept accumulated in a pollen bank. Asynchrony in flowering will no longer be a problem. The frequency of cumulative pollinations could be increased, which will result in more number of seeds set. In capsicum, pollen fails to effect normal fertilization after 3 days of anthesis. The dry and powdery pollen grains of capsicum are extremely susceptible to even mild variation in temperature and humidity. Pollen desiccates very rapidly under field conditions and do not germinate at all if collected 2 hours after dehiscence. So, if pollen is collected and stored immediately after anthesis at -1960C, the viability can be preserved and extended over a period of time. Since many of these species are widely used as source of resistance, a pollen cryobank for these species will increase breeding efficiency in capsicum.

ENABLING EX-SITU CONSERVATION OF LANDRACE FIELD

Breeders have a renewed interest in landraces of crop species (Ehdaie and Waines, 1989) since they usually exhibit useful genotypic variation for tolerance to biotic and abiotic stresses prevailing under environmental conditions where they are grown (Chang, 1985). Usefulness of the genotypic variation for local vegetable cultivars and landraces in Egypt has been the subject of studies by many researchers (Ahmed, 1996; Hussein, 1994; Khalil, 1996; Mohamed, 1996). Ryder (1991) used the lettuce (lactuca sativa L.) accession 251245 introduced from Egypt as a source for resistance to lettuce mosaic virus in breeding program for the lettuce cv. 'Salinas 88'.

Substantial amount of green pepper (Capsicum annuum L.) in southern Egypt is still produced using seeds of a highly heterogeneous local cultivar. The cultivar is highly variable for fruit size, shape and pungency. However, it is favorable for the local consumers who use its fruits both as fresh vegetable and as pickle. High per cent (up to 70%) of outcross may occur in pepper (George, 1985; Todorov and Csillery, 1990). This local cultivar, therefore, could be considered mixture of heterozygous genotypes originated in the past from varietal intercrosses that occurred in the grower's farms where they produce their own seeds without adequate isolation and no practising for plant roguing. In southern Egypt, this cultivar is usually produced under conditions of low farming inputs, interrupted schedules of irrigation, and temperatures higher than optimum for production of pepper. Therefore, ex situ conservation of such landraces may be a useful tool to maintain breeding resources for improvement of the pepper.

Towards enabling ex situ conservation, the bjective of this study, therefore, was to investigate the response of shoot-buds exised from mature plant of landrace field collections to cloning and multiplication in vitro.

Materials and Methods

Shoot-buds in leaf axils of the basal 5 to 7 stems node were excised from field grown mature plants (at the end of the growing season) of Egyptian pepper landrace (Capsicum annuum L.) collections grown in the Agricultural Research Station, Assiut University, Egypt. These shoot-bud explants were stirred for 10 sec in 70% ethanol followed by 10 min in 0.5% sodium hypochlorite (10% Clorox plus two drops of Tween-20 per liter). Then they

were rinsed three times with sterile distilled water. The disinfested explants were cultured on MS (Murashige and Skoog, 1962) medium under aseptic conditions. The medium contained 3% sucrose and 0.7% agar. The following supplemented of benzyladenine (BA) and Indole-3-acetic acid (IAA) were used (in μM): 1) 1 BA alone or plus 0.3 IAA, 2) 3 BA alone or plus 0.9 IAA, and 3) 5 BA alone or plus 1.5 IAA. The medium without these supplements was used as reference treatment. The pH of the medium was adjusted to 5.7 before being autoclaved (1.4 kg.cm-2 at 121°C).

The seven treatments were arranged in randomized complete-blocks with four replicates. Each replicate contained ten culture vessels from each treatment. The axenic cultures of the explants were kept at 25°C±0.5 under light (16h/day) from cool-white fluorescent tubes. Two consecutive experiments were conducted using randomly sampled plants within the field grown landrace plants. Data were recorded after 4 weeks of the culture on percentage of explants developing shoots and number of shoots per cultured bud. Individual shoots were separated from the multiple shoot-buds developed on primary medium containing BA alone or plus IAA. They were pooled and randomly assigned to culture on the medium either lacking or containing 2 μM a-naphthaleneacetic acid (NAA). Single shoots developed in the primary cultures from the control treatment were kept separate on the medium lacking or containing 2 μM NAA. Data were recorded after 4 weeks on the number of shoots forming roots. The rooted shoot-buds were transplanted into 5-cm plastic pots containing sterile mixture of equal volumes from sphagnum peat, washed sand, and soil. These pot cultures were acclimatized under plastic chamber. Two weeks later, acclimatized plantlets were transplanted in the plastic house. All data of the study were analyzed according to the statistical methods explained by Gomez and Gomez (1984).

Results and Discussion

No multiple shoot-buds were observed in the explant cultures on the medium lacking BA alone or plus IAA. Single shoot developed in 60 to 70 per cent of these cultures 3 to 4 weeks after incubation. Forty to fifty per cent of the developed single shoots had roots. The remaining shoots formed roots after additional 3 to 4 weeks when they were transferred to the medium containing 2 μM NAA. Multiple shoot-buds developed in cultures on the medium supplemented with BA alone or plus IAA. Seventy to eighty per cent of these cultures had multiple shoot-buds. Greater number of shoots-buds were obtained from cultures on primary medium containing 3 or 5 μM BA with or without IAA than those which incubated on the medium with 1 μM BA alone or plus IAA. Proliferation of callus tissues associated with the shoot-bud development in all cultures when IAA was added to the medium containing BA. However, callus proliferation did not affect the per cent of the explants forming multiple shoot-buds and the number of shoot-buds produced in these cultures.

Sun and Wang (1990) reported that the optimal medium for shoot-bud multiplication contained 1 mg/L BA (i.e., 4.4 μM) plus 0.5 mg/L IAA and GA (i.e., 2.85 and 1.44 μM, respectively). Also, IAA stimulated shoot-bud regeneration from different seedling explants in 9 cvs./lines of chili pepper. However, Ebida and Hu (1993) regenerated the greatest number of shoot-buds from shoot tip explants on BA containing medium with no auxin supplements. They observed only callus and roots when 0.4 mg/L of NAA (i.e., 2.15 μM) was added to the BA culture medium. Genotypic differences and/or explant type affect the response of pepper cultures in vitro.

SOMATIC EMBRYOS OF CAPSICUM ANNUUM L., GENETIC SPECIALITIES OF FORMATION

It is well known that the process of mass production of pepper regenerants, in vitro is far from its complete understanding. The main difficulties, researchers face with, are the follows: low output of regenerants per explant or absence of regeneration at all, weak repetition of the results, low rooting capability during passaging in culture. It is true for practically all in vitro directions but especially actual for organogenesis and somatic embryogenesis. In this connection according to our point of view, not empirical elaboration of protocols of different methods for concrete genotypes is perspective but fundamental researches connected with genetics of biological development of studied object and gene expression. In the case of somatic embryogenesis of pepper the following strategy of research is suggested:

1. Estimation of gene bank of studied species in order to single out different donors of a trait;

2. Detailed characteristic of trait expression during ontogeny;
3. Revealing biotic and abiotic factors, which modulate expression;
4. Studying laws of variability and inheritance of a feature in different types of donors revealing possible pleiotropic effects;
5. Working out methods of controlling expression of a character in the necessary direction.

Systemic approach to solution of the problem of formation somatic embryos lets to make a prognosis on probability of presence a gene family in population of a species and responsiveness of defined genotype within studied character. Also it lets to optimise media for concrete type and finally not only the expression within the frame of reaction norm of the genotype, but qualitative change it.

Earlier (Timina et. al. 2001) we identified different donors of formation of somatic embryos of pepper. The goal of given work was to study the expression and inheritance of the character in some donors during ontogeny.

Materials and Methods

Studied gene pool of Capsicum annuum L. was presented by varieties and lines of var. annuum, according general classification of the genus Capsicum L. (Baral, Bosland, 2002) and 35 genotypes, among them 12 F1 hybrids and 1 interspecific with C frutescent L were compared. We studied formation of general quantity of embryos, development of the embryos up to the stage of tulip, dividing received structures into normal (green) and chimerical (albino). Immature zygotic embryos on the different stages of development were used as explant. We also used other types of explants: hypocotyls, cotyledons and leaves from plants, which were sterile grown on the medium containing ingredients according Murashige and Skoog double diluted, without hormonal additions. The most responsive genotypes were used in semidiallel crosses, for studying of inheritance of different stages of embryogenesis. Plants were grown in plastic unheated greenhouse during spring-summer rotation according to general technologies. Pollination of maternal form was done by redundancy of paternal form. Embryos of parents and F1 hybrids were excised aseptically on all stages of their development: globular, heart, torpedo, stick, semi-ring and ring and then with other explants raised on medium supplemented by 2.4 D according Binzel et al., 1996 in double repetition. In each repetition at least 10 tubes with defined type of explant were used. Data were worked out using the degree of dominance hp and genetic parameters of diallel crosses.

Results and Discussion

Expression of Formation of Somatic Embryos in Several Donors

Screening of C. annuum gene bank carried out earlier has made it possible to reveal donors of both direct and indirect embryogenesis (Timina et. al., 2001). The exact timing of induction of somatic embryos connected with concrete stage of explant development in ontogeny (torpedo and stick) has been the distinctive feature of direct embryogenesis of studied donors and F1 hybrids received on their ground.

Depending on genetic background, responsive ones are explants even on the stage of globe and heart, that we could see studying F1 of interspecific hybrid C. frutescens x C. annuum. Probably the data received are evidence of different embryo's adaptation ability depending on genotype and its stage of development and is the result of its different surviving capacity in artificial medium. Different embryo's surviving capacity probably could become a basis for working out express-methods of adaptation ability to abiotic stressors and connected pleiotropic gene effects.

Data've been received indicate that the process of switching on the sporophytic way of embryos development is the irreversible one the case of direct embryogenesis using 2,4D as the inductor in the favourable in vitro conditions thanks to what we suppose there is a special gene system in C. annuum sooner referring to gene regulator one. The work of this supposed gene is highly depended on physiological stage of explant's donor.

Thus, explants that were excised from fruits at the beginning of technical ripeness were characterized by most intensive formation of cluster bipolar structures. Explants from plants at the full technical ripeness - by solid decreasing of induction process and explants from plants at the ending of vegetation - mainly by not embryogenic callus.

Ageing of an organism and changing of an endogenous balance of hormones-inductors on the one hand and also formation of enzymes-repressors could be the cause of such modulation of expression. Use of zygotic embryo explants at the stage of semi-ring and ring induced among majority of responsive genotypes only callus formation. At the same time there was one cross combination F1 Kolobok x L-48, explants of which on the same stage of development formed embryos directly. Received results prove data of Harini, Lakshmi Sita (Fari, 1995), who had watched the analogical process among formed mature zygotic embryo of the variety of California Wonder and probably the results gave an evidence of another mechanism of cascade activation of somatic embryogenesis of pepper.

Indirect formation of embryos was watched using donors L-49, Gogoshari and Kolobok. Callus was received from seeds with damaged seed peal, which contained both endosperm and embryo (semi-ring, ring). In three months of cultivation, callus was subcultivated on fresh medium with the same concentration of hormone. The culture was raised for two weeks illuminated with intensity of 3000 Lk and 16-hour photoperiod and then in darkness at the temperature 25°C during one month.

Thus, necessity of dedifferentiation and subcultivation and also stress impact of callus are specialities of indirect embryogenesis of C. annuum.

Inheriting Stages of Formation of Somatic Embryos

Disconnection and different formation of somatic embryos among studied donors were indirect evidence of that in vitro, embryogenesis could be divided into a number of stages:

induction, formation general quantity of embryos and also albinos and green structures, being taking under account by us at the stage of tulip. And the induction capability was a characteristic for the process as a whole, and other stages were its details.

Defined degree of dominancy of F1 hybrids changed differently in dependence on cross combination or genetic background and stage of embryogenesis. In general, though for the whole process we indicated prevalence of over negative and negative dominancies, the stage of general quantity of embryos was characterized by total prevalence of over positive and intermediate dominancies. The stage of albino forming was characterized by total equality of negative and positive dominancies and the stage of green structure formatting - by over expressed negative dominancy. Such differentiation gave an evidence of that each stage of embryogenesis was controlled by independent genetic systems. There were the presence of cross combinations that had hp in absolute figures higher than 1 that was an evidence of epistatic types of genetic interactions and of different genetic systems controlling stages of embryogenesis of studied donors. It is interesting to mark one more type of genetic interactions characterizing 4th stage of embryogenesis - complementary or new formation. Besides these are most valuable identified gene systems stipulating further regeneration of plants, they are different at donors L-48, L49 and variety of Prometey as we can understand from the functional test of allelism.

Found gene interactions did not let to define genetic parameters on the basis of semidiallel crosses because of not matching of experimental material to suggested model and this was confirmed by insignificant

coefficients of regression of each stage. That is why for more detailed genetic identification of donors, in future, genetic analysis using of F2, F3 and becrosses will be fulfilled. Repetition of genetic analysis of F or the same combinations but with using explants on the stage of stick showed for number of combinations complete decreasing of formatting of somatic embryos, changing of types of gene interaction and dominancy, which stipulated different expression of the trait during ontogeny. Thus, investigations have been fulfilled by us give evidence of the presence of at least two types of gene regulation of formation of somatic embryos of C. annuum. The first one is connected with the functioning of supposed gene regulation switching on the sporophytic way of development on the earliest phases of ontogeny of immature zygotic embryo.

The second - is connected with changes of gene interaction and the degree of dominancy functioning on later phases of ontogeny. Probably the first type is stipulated by the influence of inductor during transcription, the second type - during processing, cotranslational transfer or translation. Although we have not known yet the mechanisms of cascade activation and repression of the process of somatic embryogenesis in vitro, but found specialities of expression and influence of gene background let us work out aimed decisions for improving of expression of the trait by synthetic selection just today.

11

Effects of Cryopreservation of Pepper and Eggplant Seeds

INTRODUCTION

Cryopreservation at -196 °C in liquid nitrogen (LN2) can be used to conserve reproductive plant material and minimize its deterioration in storage, and has been successfully employed for seeds, ovaries, dormant buds, pollen, meristems, tissue cultures, embryonic axes, somatic embryos, excised embryos, etc. Results have been encouraging for conserving orthodox seed of cultivated and wild plants, and even recalcitrant seeds. Several factors have been shown to exert a positive, although somewhat variable influence on the outcome; these are seed moisture content, lipid content, cooling rate, length of time under LN2, and conditioning, after thawing but before germination, of recalcitrant seeds. This paper concludes a series of tests conducted at the Vegetable Germplasm Bank, University of Turin, on pepper and eggplant, to investigate factors such as seed moisture, cooling rate, and cycles of freezing and thawing.

MATERIALS AND METHODS

Four experiments were conducted, to analyse the effects of seed moisture, number of cooling and warming cycles, and cooling rate.

The seeds used were those of pepper (Capsicum annuum L.) cv. 'Corno di toro' and eggplant (Solanum melongena L.) cv. 'Prospera'. Control samples were held at - 20 °C in our germplasm bank. Samples to be cooled in LN2 were first placed in plastic cryovials.

Each freeze-thaw cycle (cooling to - 196 °C, warming to ambient temperature, and return to LN2) was either rapid (from LN2 to + 30 °C for 15 min, ambient temperature for 60 min, return to LN2) or slow (from LN2 to - 20 °C for 60 min, - 10 °C for 60 min, 0 °C for 60 min, ambient temperature for 60 min, and direct return to LN2).

Seed viability after treatment was checked using 8 replicates of 100 seeds each, performed according to International Seed Testing Association rules (1999), at 20 °C for 16 h/d plus 30 °C for 8 h/d.

In experiments 1 and 2, seeds germinated each day up to the fourteenth day were counted, and we determined percentage germination (PG), mean germination time (MGT) and percentage of abnormal seedlings (AS). In later experiments, attention was focussed on seed vigour rather than on germinability. Thus in experiment 3 an ageing test was introduced such that all seed, after the freezing treatments, was held for 48 h at 40 °C and 20 % relative humidity, before the germination tests. In experiment 4, a controlled deterioration test was similarly introduced. This consisted of a period of 48 h at 45 °C and 28 % RH (S. melongena) or 24 h at 45 °C and 18 % RH (C. annuum).

Results

Experiment 1 showed that the two methods of cooling and warming (slow or rapid) gave similar results. Thus in the later experiments only the rapid method was used. The number of cycles (from 0 to 25) had a statistically significant effect on PG in both pepper and eggplant. However, even the maximum number of cycles had relatively little effect on PG (3-4 %). In addition, unexpectedly, the lowest mean PG values (80.5 % versus 86.7 control) for pepper were found for the seeds with the lowest moisture content (4.7 %).

Experiment 2 showed that both the number of cycles and the moisture content did not influence PG in pepper. In contrast, for eggplant, the negative effect of greater seed moisture (10.5 %) was already significant after 10 cycles.

Experiment 3 showed the negative effect of high seed moisture content both in eggplant and pepper. In the latter, significantly more abnormal seedlings appeared at high humidity and with increasing numbers of freeze-thaw cycles, with synergism between these two factors. Above 15 % RH, the viability of pepper seeds was strongly depressed by numerous freeze-thaw cycles.

Experiment 4 also showed that high seed moisture was the main factor in reducing PG. In pepper, the number of cycles and the interaction number of cycles x seed moisture content also had a significant effect.

The two kinds of seed tested proved to be equally resistant to the changes in temperature consequent upon interruption of cryoconservation. Unexpectedly, different rates of cooling did not appreciably affect subsequent germinability, and at least at low values of seed moisture content, the negative effects of cooling and rewarming on viability were relatively slight.

Damage sporadically appeared in seeds with the lowest RH. This could mean that survival under cryoconservation and resistance to repeated thermal cycling, which were normally high, could be reduced by strong initial dehydration used as a preliminary to conservation. Our results also showed that, at least for the dryest seeds, cell damage consequent on numerous freeze-thaw cycles was minimal.

SCREENING OF 69 HOT PEPPER LINES FOR RESISTANCE AGAINST CUCUMBER MOSAIC VIRUS BY MECHANICAL INOCULATION

Planting of CMV resistance cultivars is the most effective control measure against virus infection in hot pepper (Green and Kim, 1994) and especially in Indonesia since strains of CMV and insect vector spreading it always exist in the field. Therefore, breeding for CMV resistance hot pepper is necessary and identification of donor lines carrying CMV resistance characters need to be conducted. The objectives of this study were to evaluate response of 69 hot pepper lines against the most severe CMV strain reported in Indonesia and to identify the resistance lines.

Severe isolate of CMV from Indonesia (CMV-02) was obtained from Research Institute for Vegetable, Lembang, Indonesia and maintained by mechanical inoculation in tobacco var. Xanthii. For hot pepper evaluation, CMV was prepared by extracting leaf sap of virus infected tobacco showing symptoms of CMV infection. For each of 69 lines, 25 seedlings grown individually in 200 ml plastic pot containing sterilized potting mix were evaluated. Seedlings were maintained under an insect proof glasshouse and were mechanically inoculated twice with CMV-02 at cotyledon and two leaf stage, respectively. The inoculated cotyledons and leaves were dusted with carborundum (600 mesh) prior to rubbing the sap of CMV infected tobacco.

Number of seedlings showing symptoms, period of incubation, and types of symptoms of CMV infection in the inoculated seedlings was observed weekly up to four weeks after the second inoculation. Representative of symptomatic and all symptomless seedlings were verified for the presence of CMV particles using DAS-ELISA (AGDIA). Disease intensity (DI) was calculated using the following equation 1) This research was supported by Graduate Tim Research Grant, URGE Project, Ministry of National Education, Republic of INDONESIA. with n, v, N, and V were number of seedlings with symptoms, score of symptom severity in individual seedlings, number of seedlings evaluated and the highest score (5) of symptom severity, respectively. Symptom severity was scored as follow: score 0: no symptoms observed; score 1: necrotic local lesion, no systemic invasion or very mild chlorosis; score 2: mild to moderate mottle or mosaic without significant leaf distortion; score 3: moderate to severe mosaic with significant leaf deformation; score 4: severe mosaic with severe stunting and leaf deformation; and score 5: very severe mosaic, leaf deformation, and stunting. Subsequently, the tested lines were grouped into six response groups based on the calculated DI and results of DAS-ELISA Results and Discussions The calculated disease intensity, based on score of symptoms in the inoculated seedlings, among 69 hot pepper lines ranged from 0 to 98%. Four lines (C1024, KA-2, PBC495, and C1042) showed DI = 0 while three lines (LV2323, Tit Paris, and PBC068) showed DI

Mechanical inoculation of CMV on 69 hot pepper lines resulted in ranged of response from symptomless to severe mosaic and leaf deformation. Lines grouped into moderately to highly susceptible showed 100% symptomatic seedlings after inoculation with CMV. For seedlings showing symptoms, the incubation period ranged from 8 to 26 days after inoculation. Lines with shorter incubation period (IP) showed severe mosaic and leaf deformation while one with longer IP showed mild mosaic or mottle symptoms.

Some of the moderately resistance and all resistance lines were segregated for symptomless and symptomatic seedlings after inoculation with CMV. Symptomless seedlings remained symptomless after reinoculation with CMV but most of them showed positive results of DAS-ELISA.

< 10%. The rest of the lines evaluated showed DI ranged from 11 to 98%. The C1024, KA-2, PBC495, and C1042 lines showed 100% symptomless seedlings after twice inoculation with CMV. Moreover, DAS-ELISA using leaf sap showed negative results indicating the absence of CMV particle in the inoculated seedlings. The tested seedlings remained symptomless and gave negative results of DAS-ELISA after reinoculation with CMV. Identification of hot pepper lines with resistance to local strain of CMV is necessary for breeding program. In this evaluation we have been able to identify four different lines showing immunity to CMV. Since the selected lines is immune against the most severe isolates of CMV from Indonesia, they may be used to overcome major problem associated with severe CMV infection in the field. Although screening of hot pepper lines against CMV have been conducted, none reported lines with resistance character against CMV such as the one identified in this report. References Duriat AS, 1996. Management of peer viruses in Indonesia: Problem and Progress. Indon. Agric. Res. Dev. (IARD) J. 18: 45-50 DI= S (n x v) N x V x 100%,

In this evaluation we have identified four hot pepper lines that were symptomless and showed negative results for DAS-ELISA. According to Dolores (1996), such lines were classified as immune against CMV. However, further testing is needed whether these lines were actually immune, unable to support replication, or unable to support systemic spread of CMV.

These four lines can be used as donor parents for developing CMV resistance hot pepper varieties. Studies on the genetics factors governing the resistance character will be beneficial for setting up breeding program for CMV resistance cultivars. Moreover, C1024 and C1042 were originated from Indonesia, PBC495 was from India, and KA-2 was from Sri Langka, respectively. It is possible that some of these lines carry different resistance mechanisms against CMV. In such cases, the different resistance mechanisms can be pyramided into single genotype to develop more durable CMV resistance hot pepper cultivars.

RED PEPPER (PAPRIKA) VARIETIES RESISTANT TO BACTERIUM AND THEIR ROLE IN CULTIVATION

The gene stock of the population of pepper developed in the Carpathian basin, as a result of long selection work, proved to be suitable to make Hungarian pepper (paprika) world famous. Producer and consumer requirements with pepper varieties necessitated the expansion of the gene stock of Hungarian varieties. Incorporation of genes ensuring resistance against bacterium disease of pepper lasting almost a decade has also set this aim.

Under the Hungarian conditions only the Bs-2 gene causing hypersensitive reaction, coming from Capsicum chacoense wild species and the gds (general defense system) gene redeveloping the general defense system of the host plant found in C. annuum PI. 163 192 line ensure resistance against Xanthomonas vesicatoria bacterium.

Resistance improvement of vegetable plants so far has been based on reaction concomitant with fast tissue destruction following the entry of pathogens. In course of resistance improvement of eating pepper varieties against bacterium, Bs-2 gene ensuring satisfactory resistance level was built into spice pepper (paprika) varieties, too.

Among the undesirable, hardly penetrable characteristics, closely connected to the gene ensuring hypersensitive reaction, primarily small fruit size delayed the production of resistance types. After almost a decade long work it was possible to unite the adequate resistance with economic characteristics required from pepper varieties. Kaldom and Kalorez varieties candidates represent this result.

The other trend of our resistance improvement program is the incorporation of the gene, the gds gene we described, localizing the pathogen with entirely different strategy than the hypersensitive reaction.

The reaction determined by the gds gene is aimed at keeping alive the cells affected by the pathogen by all means. The reaction based on cell growth and cell wall thickening provides satisfactory defense both against pathogen bacterium and fungus species and against environmental effects. In addition, it provides great growth strength for the varieties.

As a result of our work, we plan a significant role for varieties resistant to Xanthomonas vesicatoria bacterium in making Hungarian spice pepper (paprika) cultivation profitable. Saving the costs of repeated and inefficient chemical defense against the bacterium, itself ensures reasonable profit for the producers. We have also taken into consideration that disease resistant varieties might become fundamental elements of organic cultivation.

SCREENING OF SWEET PEPPER GERMPLASM FOR RESISTANCE TO BACTERIAL WILT (RALSTONIA SOLANACEARUM)

Bacterial wilt in capsicum caused by Ralstonia solanacearum E.F. Smith has become a serious problem in India. The commercial varieties are susceptible to this disease and chemical control through treatment of soil is cumbersome and uneconomical. That is why, breeding varieties for bacterial wilt resistance combined with high yields and acceptable quality is the present day need. Occurrence of this disease is associated with high temperature (above 32°C) and sufficient moisture in the soil. Yield losses up to 100 are reported in wilt prone areas. Keeping this in view an investigation was undertaken to test a wide range of germplasm collection of sweet pepper, with the objective that they can either be utilised directly for cultivation to be find out germplasm sources of resistance to bacterial wilt to be utilised ultimately in breeding resistant cultivars.

Materials and Methods

A set of thirty genotypes of sweet pepper comprising exotic and indigenous genetic stock collections as well as a few commercial lines with good performance was planted in a completely randomised block design with three replications. The disease intensity was recorded under natural sick plots maintained at the vegetable experiment farm of the Department, under field conditions.

The wilting of the susceptible check indicated the presence of virulent inoculum in the soil. Bacterial ooze test was carried out on all the wilted plants to confirm bacterial wilt. The disease rating was done as per the scale suggested by Mew and Ho (1976).

Resistant: < 20 wilting

Moderately resistant: 20 to 40 wilting

Moderately susceptible : 41 to 60 wilting

Susceptible : > 60 wilting

Results and Discussion

The genotypes IHR-546 and PBC-631 were highly wilt resistant genotypes of capsicum. The lines Arka Gaurav, PBG 5005, Cap B, Bastidon and Cap C were observed to be moderately resistant and remaining entries ranged from moderately susceptible to susceptible disease attack against bacterial wilt. Other workers have also reported the resistance against bacterial wilt in sweet pepper. The resistance has been observed by Matsunaga et al., 1993 and Wang et al., (1997).

The wilt resistant accession, IHR-546 was dark green fruited, compact growth habit with pungent fruits. Fruit length is about 6-7 cm. Line PBC-631 had long fruits like paprika type, light green coloured and sweet in taste. Lines Cap B and Cap C (moderately resistant) were having erect conical type of fruits. Bastidon had normal bell shaped fruits.

The present studies indicate that the resistant gene for bacterial wilt is available in the capsicum strains and same may be incorporated into otherwise suitable commercial cultivars. Therefore, a breeding programme involving genotypes viz., IHR-546 and PBC-631 can be envisaged to transfer bacterial wilt resistance in a single genotype coupled with high yield potential.

Development of Resistant Pepper Lines against Anthracnose using Interspecific Crossing between Capsicum Baccatum and C. Annuum

Fruit rot of chili pepper caused by anthracnose (Colletotricum gloeosporioides) is one of the most serious diseases in Korea. Resistance was identified only in different species, Capsicum baccatum and C. chinensis, by Pae et al. (1995) and Black (1998), respectively. Unfortunately, these two species cannot be easily crossed with C. annuum. In order to overcome interspecific crossing barrier and to restore agronomic characteristics at the levels of commercial chili pepper, we used several breeding strategies including embryo rescue, backcross, and advancement of generation based on phenotype selection. We report the results of embryo rescue, and the performance of some lines, resistant to anthracnose with good agronomic characteristics.

Materials and Methods

Interspecific cross & embryo rescue: Four hundred and fifteen seeds of 62 days old immature fruits derived from reciprocal crosses between C. baccatum, 'UK1', and C. annuum, 'L.F.', were sown on 1/2 MS medium under conditions of 25_ and 12hr day length. Seedlings were transplanted in greenhouse 14 days after germination in 1/2 MS medium.

Bioassay: A Colletotricum gloeosporioides isolate was prepared from infected fruits collected in Korea on potato dextrose agar (PDA) medium by tissue isolation technique. Inoculum of 5 x 105 spores/ml was sprayed with 1.5kg/cm2 pressure. The anthracnose resistance was investigated in parental cultivars and two hundred and twenty three progeny plants derived from single-, three way-, and back-crosses. Five fruits from each plant were inoculated and kept at 25±2_ in moist chamber. The disease symptoms were recorded seven days after inoculation. The degree of disease incidence on fruit was judged by visual observations, using the scale 1 to 9. Plants falling in scale 1 to 3 were determined as resistant.

Results and Discussion

F1 seeds derived from C. baccatum x C. annuum ('UK1' x 'L.F.') showed 4.6% survival rate in greenhouse condition. Through embryo rescue in 1/2 MS medium, we finally obtained nine F1 plants, which were highly sterile. However, the reciprocal crosses did not produce any survival plant. These results were similar with the previous report.

Progenies from three-way crosses involving C. annuum as either female or male parent showed very low fertility, i.e., 10.5 - 14% in F1. On the contrary, F2 generation from backcrosses gave reasonably high rate of fertility, 29.7 and 73.5%. In F1 generation, most of the fertile plants were resistance to anthracnose, whereas, half of F2 and F4 plants were susceptible.

Fourteen were selected from one hundred and thirty progenies as having the high degree of resistance to anthracnose with good fruit characteristics. The fruits of these lines were similar with C. annuum ('LF') in length and width of fruits, whereas the weight of fruits had an intermediate value.

Only one report has been published on interspecific hybridization between C. baccatum and C. annuum in relation to disease resistance. Dumas de Vaulx and Pitrat (1977) reported resistance of PVY and Phytophthora capsici in F2 progenies of C. baccatum and C. annuum, but so far no report has been found on the resistance to anthracnose.

EVALUATION OF CAPSICUM SPP. GENOTYPES FOR RESISTANCE TO PHYTOPHTHORA CAPSICI IN BRAZIL

Phytophthora wilt is the most important fungus disease of Capsicum spp. in Brazil, particularly destructive under high humidity and high temperature. Most of the Brazilian cultivars and hybrids are susceptible to Phytophthora capsici and resistance to phytophthora wilt is important for Capsicum breeding programs. The use of resistant cultivars is the most effective method of disease control. Chemical control is relatively efficient but is expensive and can cause genetic mutations on the pathogen.

For the last twenty years, the Embrapa Hortaliças Capsicum breeding program has been working with multiple resistance and some of the released resistant lines are been used in Brazil and other countries. The objective of this work was to evaluate part of the Embrapa Hortaliças Capsicum spp. germplasm collection for resistance to phytophthora wilt in order to identify news resistant genotypes.

Materials and Methods

During the period from 1998 to 2001, 363 Capsicum genotypes of the germplasm collection (245 C. annuum, 42 C. baccatum, 36 C. chinense, 28 C. frutescens and 12 wild Capsicum) were evaluated to phytophthora wilt. Four different experiments were carried out under greenhouse conditions in a randomized experimental design with three replications (7 plantlets/replication). Plantlets were inoculated 40 days after emergence by placing 3 ml of 5 x 104 P. capsici zoospore suspension per plant of the CNPH 08 isolate (Reifschneider et al., 1986). Genotypes CNPH 148 and 192 were used as resistant and susceptible checks, respectively. Evaluations were made 7 and 15 days after inoculation by determining the percentage of dead plants.

Results and Discussion

Among the genotypes tested, those showing plantlet survival ranging from 75 to 100% were rated as resistant. Only 2.7%

of the genotypes evaluated were identified as resistant sources to phytophthora wilt, nine genotypes of Capsicum annuum and one single accession of the C. parviflorum. The susceptible (CNPH 0192) and resistant (CNPH 0148) genotypes were included in all four experiments and performed as expected. The determination of new resistance sources and the introgression of Phytophthora capsici resistance genes in other domesticated species is strategic to the development of C. baccatum, C. chinense and C. frutescens resistant cultivars. This method is particularly important for those materials that do not have satisfactory resistance level.

Embrapa Hortaliças Capsicum collection is one of the largest in Brazil, and has more than 1,200 accessions, mostly of C. annuum, C. baccatum, C. chinense and C. frutescens. Additionally, ten wild Capsicum species recently founded in the remains of the Atlantic forest in Brazil were characterized and maintained at the germplasm bank. The search for new sources of resistance to Phytophthora capsici in the Capsicum germplasm collection of Embrapa Hortaliças will continue to test new entries, especially wild and semidomesticated species collected in Brazil, since disease resistance is one of the most important goals of the Embrapa Hortaliças Capsicum breeding program.

ANALYSIS OF DNA METHYLATION DURING GERMINATION OF PEPPER SEEDS BASED ON METHYLATION-SENSITIVE AFLP MARKERS

DNA methylation in plants is related to a number of epigenetic phenomena among which the regulatory mechanisms during development and differentiation.

Here we report our preliminary results on the analysis of cytosine methylation during pepper seed germination using MSAP (methylation-sensitive amplified polymorphism), an adaptation of the AFLP (amplification fragment length polymorphism) technique.

The technique is based use of the isoschizomers HpaII and MspI, that differ in their sensitivity to methylation of their recognition sequences. Both enzymes recognize the tetranucleotide sequence 5'-CCGG, but their action is affected by the methylation state of the external or internal cytosine residues. HpaII is inactive when either or both of the two cytosines is fully methylated (both strands methylated) but cleaves the hemimethylated sequence (only one strand methylated), while MspI cleaves hemi or fully methylated C5mCGG but not 5mCCGG.

Material and Methods

DNA was extracted every 48 h from embryo tissues of germinating F1 pepper seeds (cv. "Corno di Toro"). We used two consecutive PCRs to selectively amplify the EcoRI-HpaII (H) and EcoRI-MspI (M) DNA fragments. The pre-selective amplification (first PCR) was performed using EcoRI and HpaII/MspI adapter-directed primer, each possessing a single selective base (E+A; HM+T).

Within each AFLP primer combination patterns obtained in 10 lanes were compared, corresponding to five different sources of pepper genomic DNA (i.e. from dry seeds and seeds germinated for 2, 4, 6 and 8 days) each restricted with one of the pairs of enzymes EcoRI/HpaII (lane H) or EcoRI/MspI (lane M).

Results and Discussion

Comparison of the AFLP patterns revealed two main kinds of polymorphism: (i) bands always appearing after digestion with H but not with M or vice versa (i.e. 131 out of 936); these bands were the result of DNA methylation at the 5'-CCGG sites but did not highlight changes in methylation pattern during germination (Figure 1a, Table 1), (ii) bands showing polymorphism at different stages of germination (i.e. 328 out of 936); these bands resulted from changes in DNA methylation status.

Changes in DNA methylation status were mainly displayed as: (i) fragments not found in dry seeds but present after digestion with H and M at a certain stage during germination, i.e. 215 out of the 328 polymorphic bands; (ii) fragments present after digestion with H and M in dry seeds but no longer detected at a certain stage during germination, i.e. 44 out of 328 polymorphic bands.

Demethylation preferentially originates short fragments, as a previously methylated 5'-CCGG recognition site near an EcoRI recognition site, can come within reach of MspI or HpaII or both enzymes. As reported by Xu et al. (2000) and Reyna-Lopez et al. (1997), short DNA fragments, ranging from 50 to 1500 bp, can be more effectively amplified by AFLP or detected in a sequencing gel.

During germination, the observed high frequency of fragments originated de novo (i.e. 215) and the concomitant disappearance of fragments which were detected in dry seeds (i.e. 44) may thus be attributed to demethylation events. Presumably, the fragments originated de novo were observed at higher frequency since, being shorter, they had a higher chance of amplification and detection. The hypothesis that we were detecting demethylation events is further supported by the fact that, during germination, after digestion with both MspI and HpaII, we observed a progressive increase in the frequency of fragments detected together with a progressive decrease in the frequency of fragments no longer detected. By the same interpretation, the appearance at higher frequency and the disappearance at lower frequency of some fragments, after digestion with only one of the two isoschizomers, might also be attributed to demethylation events occurring during germination.

Interestingly 49 of the 215 fragments detected at a certain stage of germination underwent a second change in methylation status, since they were no longer observed in a subsequent phase of germination. This may be further on interpreted as the occurrence of a second demethylation event, originating shorter fragments which were differently positioned in the sequencing gels. The interpretation of our data, as reflecting progressive DNA demethylation occurring during germination, is supported by the fact that in plant nuclear genomes cytosine methylation is known to play an important role as regulator of gene expression in development, and methylation in particular regions of genes or in their vicinity can inhibit the expression of these genes, while artificial demethylation of genes is known to result in reactivation.

The results here reported show this technique to be highly efficient for large scale detection of cytosine methylation in germinating seeds. The ability to isolate and amplify these MSAP fragments may thus make possible direct identification of sequences (or genes) which play a key role at different stages of germination.

IRRIGATED AGRICULTURAL RESEARCH PROGRAM, NATIONAL CENTER FOR AGRICULTURAL RESEARCH AND TECHNOLOGY TRANSFER (NCARTT).

Eggplant (Solanum melongena L.), also known as aubergine, brinjal or guinea squash, is one of the most important vegetable crops grown in Jordan. Most of the eggplant cultivars currently grown in Jordan are hybrids. However, some eggplant landraces are still grown in many small farms due to some consumer demand. Farmers have given various local names to some landraces; "Batteiri" and "Ajamy" are the most popular local eggplant landraces. Genetic diversity in local eggplant landraces is particularly high in Jordan. These landraces are of a great value due to their wide range of variability in morphological and physiological characters. Most of local eggplant landraces has unknown characters, therefore, the aim of this study was to evaluate productivity and characterize some morphological and agronomical traits of twenty-nine accessions of local eggplant landraces.

Materials and Methods

Twenty-nine accessions of eggplant (Solanum melongena L.) landraces collected from local farmers throughout the country during 1995, and conserved in the gene bank of the National Center of the Agricultural Research and Technology Transfer (NCARTT), were used in this study. The experiment was performed in Karama station (NCARTT station in the Jordan Valley). Forty-five day old seedlings were transplanted into the field in late Sep. 2001. Plants were spaced 0.5m apart in 5m raised beds 1m apart. Black polyethylene mulch and a drip irrigation system were used. Analysis from saturated paste extracts from the experimental field and water samples from the irrigation water are included below:

Sample	EC ($dS\ m^{-1}$)	pH	Meq L^{-1}			
Ca	Mg	Na		Cl		
Soil 0-30 cm	21.7	7.6	68	123	71	154
Soil 30-60 cm	12.2	7.7	39	97	53	126
Water	2.5	8.0	5.5	6.0	11.2	10.1

The treatments were arranged in a Randomized Complete Block Design (RCBD) and replicated 3 times. Each replicate included 30 plants.

Average number of days to first flower opening and first harvest at commercial ripeness stage, growth habit (upright, intermediate and prostrate) were recorded. Four plants from each treatment in each replicate were separated into shoot and root, dried at 65oC to a constant weight to determine average shoot and root dry weights plant-1. Fruits were harvested for several times at commercial ripeness stage starting from Jan.2nd 2002 and yield, average fruit weight and average fruit number plant-1 were calculated. The following fruit and seed characters were recorded according to the eggplant descriptor published by the International Board for Plant Genetic Resources (IBPGR), 1988: average fruit length and breadth, fruit curvature, color, flesh density and taste, seed number per fruit, color and 100 seed weight.

Results and Discussion

Variations in vegetative growth parameters were observed between eggplant landraces. Growth habit was upright, intermediate and prostrate for 14, 10 and 5 accessions, respectively. Average number of days from seed sowing to first flowering ranged from 78 to 85 days and to first harvest ranged from 144 to 177 days. Wide and significant differences in shoot and root dry weights were revealed among the eggplant landraces. Average shoot dry weight plant-1 ranged from 164 g in accession 118 to 33 g in accession 123. Average root dry weight plant-1 ranged from 19 g in accession 118 to 7 g in accession 135.

Significant variations in yield of the eggplant landraces were also observed. Eggplant yield ranged from 40 ton ha-1 in accession 118 to 13 ton ha-1 in accession 123. Yield less than 20 ton ha-1 was produced by 7 accessions, between 20 and 28 ton ha-1 by 11 accessions and between 30 and 40 ton ha-1 by 11 accessions. Average fruit number per plant ranged from 27 to 10 in accessions 121 and 123, respectively. Eleven accessions had average fruit number less than 15, ten accessions between 15 and 18 and 8 accessions between 20 and 27 fruit plant-1. Average fruit length ranged from 22 to 8 cm in accessions 125 and 139, respectively, and fruit breadth from 11 to 3.5 cm in accessions 141 and 119, respectively. Fruit breadth to length ratio varied from 1.0 in accession 139 (fruit length as long as broad) to 0.2 (fruit length several times as long as broad) like in accession 125.

Fruit curvature varied from none to snake shaped. Fruit color ranged from purple green in accession 126 to milk white in accession 121 and to black in accessions 116, 117, 123, and 131. Fruit taste varied from excellent in accession 122 to very bitter in accession 140 and fruit flesh density from very loose in accessions 117, 127 and 141 to very dense in accessions 119, 124, 132 and 137. Seed number per fruit varied from very few in accessions 139, 140, 141 to many in 14 accessions, seed color ranged from light yellow to brown and 100 seed weight from 0.27 g in accession 136 to 0.52 g in accession 125.

Considerable differences in plant vegetative growth, yield and fruit characters were observed among the accessions. Accessions with higher yields (118, 117, 127 and 121) could be recommended for further studies. The wide range of variation in morphological and agronomical characters in the local eggplant landraces might be of a grate value for breeders according to their objectives.

POLLINATION BEHAVIOUR AND NATURAL HYBRIDIZATION IN SOLANUM MELONGENA L. TO UTILIZE THE FUNCTIONAL MALE STERILE LINE FOR HYBRID SEED PRODUCTION

The aim of this study was to investigate floral biology, pollination behaviour and natural hybridization in eggplant in main cropping season (autumn-winter) of West Bengal (eastern state of India).

The studies were conducted under field growing conditions during 2000-01. Five diverse genotypes grown in autumn-winter season (September to March) under average day and night temperature ranging from 24.8 to 33.40 C and 10.2 to 25.10 C, respectively were employed for floral biology study. The stigma receptivity was studied by fruitset method employing 5 diverse parental lines and the functional male sterile line UGA-1 MS under two different environmental conditions : winter E1(November-December under day and night temperature range of 26.2-30.70 C and 11.5-18.30 C) and rainy season E2(July-August when day and night temperature lay between 32.3-33.00 C and 25.8-26.10 C). Emasculated flowers in the 6 female parents were hand pollinated with the mixed pollens between 8.30 to 12.30 a.m. in 4 stigmal stages: one day before anthesis (S1), day of anthesis (S2), one day after anthesis (S3) and two days after anthesis (S4). Thirty randomly emasculated flower buds in 5 plants per genotype were hand pollinated in each stigmal stage in one-month span.

Influence of insect pollinators on fruit set was studied through growing 5 plants each of the 5 diverse genotypes both under 2 x 2 x 3 ft nylon net cage and unprotected condition in autumn-winter season. Frequency of natural hybridization was determined through use of three homozygous genotypes possessing dominant (pigmentation in stem, petiole, vein and fruit, P and presence of prickle in stem, petiole, vein and calyx without any pigmentation, gS) and recessive (prickleless and green, gs) marker characters. Sixty each of P + gs and gS + gs plants were planted alternatively in 4 rows in symmetrical design in two separate and isolated blocks in such a manner that the row started with P or gS plants.

IN VITRO INDUCTION OF HAPLOID IN EGGPLANT (SOLANUM MELONGENA L.)

Eggplant (Solanum melongena L.) is an important vegetable crop of Indian origin holds a coveted position in Indian sub-continent in summer and rainy season. Due to good nutritional value and its suitability to grow under a wide range of growing conditions, it has been extensively cultivated in different parts of world. Immense variability exists in this crop, which provides an opportunity to the researchers to develop plants with good quality and productivity. Recently much work has been done on development of hybrids with good yield and quality. But availability / development of homozygous / inbred line is a bottleneck in efficient exploitation of heterosis as the development of inbred takes 3-4 generations of selfing. Another alternative way is to develop homozygous lines using in vitro method such as anther or pollen culture.

Haploids produced through anther/pollen culture is of great importance in plant breeding programmes as well as in basic genetic studies. Haploid and doubled haploid lines obtained by efficient anther culture technique are now opened up new possibilities for rapid fixation of genotypes and genetic analysis. Moreover it can be an excellent support for molecular analysis and location of quantitative traits. Among the various methods available for haploid production, anther culture is the most practical, viable and widely used technique. Anther culture has been applied to many species and it is recognized that the production of haploid lines greatly assist in to subsequent production of F1 hybrids (Rotino et al. (1992), Arnison et al. (1990). However, usefulness of this approach is limited in solanaceous species because some respond very poorly to anther culture.

The culture response is affected by numerous factors, such as genotype, donor plant culture conditions, pretreatment of flower buds or anthers, stage of development of microspore, culture medium composition and culture conditions. Haploid production for practical application in breeding programme of several solanaceous crops such as tomato and potato were well documented. However, very little progress has been made in case of anther culture for haploid development in eggplant. In present study an attempt was made to standardize protocol for anther culture in eggplant to obtain haploid plants.

Materials and Methods

The experimental material comprised of three eggplant F1 hybrids. Young anthers of Solanum melongena were cultured on MS medium and GD medium containing various growth regulators like 2-4 D, BA, NAA (0.5, 1.0, 2.0 mg/l) for callus induction. The anther was crushed in acetocarmine to test the stage of pollen development.

The young isolated buds were pretreated in cold for 2-3 hrs. These buds were surface sterilized with 0.1 % HgCl2 for 3 minutes and washed thrice with sterile water. The anthers were aseptically cultured on MS as well as GD medium. Anthers alongwith their filaments were excised under aseptic condition and placed on a sterile petri dish horizontally and sealed with parafilm. The cultures were incubated in dark for initial 20-30 days at 25 ± 2 oC and after callus initiation all the cultures were transferred to light for 16 hour photoperiod. A minimum of 200 anthers from 5 plants were cultured for each treatment. After four week of incubation the anthers were observed for callus initiation. Cultures were scored for frequency of callus induction at the end of four weeks. The frequency was calculated as the ratio between the numbers of anthers responded to callus induction or regeneration to that of total number of anthers inoculated. The developed plantlets from the anther culture were transferred to small pots. After hardening (20-30 days) they were transplanted to the green house.

Results and Discussion

Culture of anthers to raise haploids followed by chromosome doubling is an easy method of developing homozygous lines in relatively less time than traditional methods. In eggplant callus initiation from anther has been reported earlier by Vaulx and Chambonnet (1982), Yadav et al. (1989) and Karakullucku and Abak (1993). However, regeneration was poor in these cases. In present study, in vitro regeneration was carried out to produce haploid through anther culture. Selection of anther/ correct stage of pollen development were found to be the most critical factor for haploid production.

EFFECT OF HOST NUTRITION ON PHOMOPSIS DISEASE OF BRINJAL

Phomopsis leaf blight and fruit rot of brinjal (Solanum melongena L.) caused by Phomopsis vexans Harter is a serious disease causing 10-25% loss in marketable fruits (Panwar et. al.1970) Although the disease has been investigated by several workers, information on the role of host nutrition, a primary component of disease control, on the disease is lacking. Since manipulation of plant nutrients play a significant role on the outcome of disease, the present study was conducted on the effect of fertilizer on the disease and crop yield.

Materials and Methods

Field trials in randomized block design with three replications on the effect of forms of nitrogenous fertilizers and different doses of nitrogenous, phosphorus and potassium fertilizers were conducted at the experimental farm of the department during 2000 and 2001 crop season with highly susceptible brinjal cv. `PPC' following recommended package of practices for vegetable crops except application of fertilizers. One-month-old seedlings of brinjal were transplanted in the first week of June during both the years with row-to-row and plant-to-plant spacing of 60 and 45 cm, respectively in 2 x 2 m2 Plots. Data on leaf blight and fruit rot were scored on 1-12 point scale with the appearance of disease and thereafter at weekly intervals till the completion of experiment. The percent disease severity was determined, pooled at the end for determining the relative effect of each treatment. To ascertain the role of forms of nitrogenous fertilizers, calcium ammonium nitrate, ammonium sulphate and urea were applied @ of 100, 102.5, 97.5 kg N, P & K /ha respectively in three splits with Phosphorus and potassium @ of 60 and 45 kg P & K /ha respectively as basal dose. Plots which did not receive N, P, K fertilizers served as check. In other field trials, Calcium ammonium nitrate (CAN) @ of 87.5, 100, 112.5 kg Nha-1, single super phosphate @ 56, 60, 64 Kg P ha-1 and murate of potash @ 30, 45, 60 kg K ha-1 with normal doses of P & K, N & K,and N & P, respectively were applied for determining the role of different doses of N, P & K fertilizers on disease and crop yield.

Results and Discussion

All nitrogenous fertilizers caused significant increase in disease severity and yield over unfertilized plots. The average severity of leaf blight and fruit rot was more in plots fertilized with urea and CAN than those with ammonium sulphate. The severity of leaf blight and fruit rot did not differ significantly in plots fertilized with urea or CAN but differed significantly with NH_4SO_4. Although, fertilization with CAN or urea significantly increased the leaf blight and fruit rot of brinjal, there existed

no significant differences in crop yield between plots fertilized with CAN and ammonium sulphate and those of ammonium sulphate and urea. Thus, application of ammonium sulphate to brinjal crop can play a significant role in the suppression of disease and in ameliorating the crop yield. A progressive increase in the severities of leaf blight and fruit rot was noticed with an increase in the levels of nitrogen over plots which received either P & K or no fertilizer. There existed no significant differences in severities of leaf blight in plots, which received 87.50 or 100 kg N ha-1. However, at these levels of nitrogen there were significant differences in the severity of fruit rot. Plots which received 112.50 kg N ha-1 showed the maximum development of leaf blight and fruit rot and differed significantly from other two doses of N fertilizer.

Leaf blight severity also differed significantly between plots fertilized with potassium, phosphorus, and those, which did not receive any fertilizer. In these plots though no difference in severity of fruit rot was observed, but there existed significant difference in crop yield. Plots, which did not receive any fertilizer, showed the minimum development of disease but with less crop yield. An increased dose of N caused non- significant effect on crop yield. Thus fertilization with lower dose (87.50 kg N ha-

1) of nitrogen will help in maintaining the disease to the minimum without any appreciable effect on crop yield. The different doses of phosphorus showed variable response on disease severity. No significant differences in severities of leaf blight were observed with an increase in levels of phosphorus. However, increased doses of phosphorus caused significant effect on fruit rot severity. Although, there was more leaf blight at 64 kg P ha-1 but had minimum fruit rot and significantly higher crop yield.

Plots which were fertilized with nitrogen and potassium, though developed comparatively more leaf blight and fruit rot but were at par in their effect on crop yield with those of 56 and 60 kg P ha-1. The plots, which did not receive any fertilizer, developed less disease but had poor crop yield. Progressive decrease in severity of leaf blight and fruit rot was noticed with an increase in doses of potassium. At 60 kg K ha-1 there was minimum leaf blight and fruit rot and maximum crop yield which differed significantly from other treatments. Plots, which received nitrogen and phosphorus, developed maximum leaf blight. However, in these plots there was less fruit rot than those plots which received 30 kg K ha-1. Although significant differences were observed in severities of leaf blight and fruit rot at all levels of K fertilizers, no differences in yield were observed between plots, which did not receive K, and those with 30 and 45 kg K ha-1. Plots with no fertilizers though developed less disease but yielded poorly.

Nitrogen applied to brinjal as calcium ammonium nitrate (CAN) and urea enhanced the severity of leaf blight and fruit rot where as nitrogen in the form of ammonium sulphate reduced it drastically. Fertilization with ammonium sulphate also supplies sulphate in addition to nitrogen and fungitoxic potential of sulphur is well documented, thereby could be less disease. These results are in agreement with Daly(1949), William(1965) who reported reduced severity of stem rust of wheat and fruit & root rot of tomato, respectively with application of NH4-N. Similarly Huber and Watson (1974) have cited several examples where disease severity decreased with ammonical nitrogen and enhanced by nitrate nitrogen and vice-versa.

An increased susceptibility of brinjal to phomopsis disease at higher nitrogen nutrition is probably due to luxuriant growth of the host thereby affecting the microclimate of crop canopy favouring the pathogen. Alternatively, it could be possible that at higher nitrogen nutrition there is enhanced exudation of nutrients which inturn enhance the inoculum potential and aggressiveness of the pathogen resulting in more disease. Huber (1980) have reported higher disease with an increased N-nutrition in Tomato- Corynebacterium michiganense, Potato-Streptomyces scabies, Potato- Alternaria solani host-patho systems. Phosphorus nutrition in majority of cases behaves like nitrogen. In the present studies, it reduced the fruit rot significantly at higher doses reduced the severity of early and late blight of potato with potassium fertilization. Potassium nutrition in general is negatively correlated with disease in many host-Patho systems. Thus application of ammonical nitrogen at recommended doses, phosphorus and potassium at higher doses will help in reducing the phomopsis leaf blight and fruit rot of brinjal and in ameliorating the crop yield especially under the Himachal Pradesh conditions. Table 1: Effect of different forms of nitrogenous fertilizer on the Phomopsis leaf blight and fruit rot of brinjal. Disease severity (%)

12

INTEGRATING SCIENCE AND MANAGEMENT OF BIODIVERSITY IN THE WESTERN GHATS

RESEARCH NEEDS FOR THE WESTERN GHATS

The Western Ghats, a chain of mountains from Gujarat to the extreme tip of India, Kannyakumari is a veritable mine of genetic resources, embedded in and described generally as flora and fauna many of which are endemic and threatened with extinction. This 1, 60,000 m^2 land mass provides the life support for the millions who live with in, in the periphery and out side. Of all the benefits that the Western Ghats offer to the mankind, the most important one, probably, is that it forms the catchments area for a complex of river system that drains almost 40% of India.

The innumerable development projects, mostly forest consuming, on the hills, lead to soil erosion, submersion of large extent of forests for river valley projects, denudation of forests, fragmentation of the forests, raising of commercial plantations, setting up of industrial establishments polluting the air, soil and water, expansion of agricultural activities, usage of tonnes of pesticides, establishment of industries, conversion of forests into commercial plantation, logging, mining, encroachment, have taken a heavy toll on the natural ecosystems of the Western Ghats. Large extent of natural forest cover was destroyed and degraded.

The remaining few stretches of forests, however, still house a biological wealth earning it the distinction of one among the

25 hotspots of the biodiversity in the world. In spite of all the development projects, social injustice and inequities could not be brought down which exacerbate the pressure on the remaining natural wealth. Therefore, efforts should be taken to formulate a comprehensive plan for the restoration, maintenance and preservation of the life support systems of the Western Ghats. Such a plan would essentially require, inter alia, the following.

Resource Documentation

(1) Detailed resource map of the entire Western Ghats, (2) a scheme to monitor the resources every two years, (3) micro level resource maps as done in some of the districts of Kerala.

Long - term Conservation

Review of the existing PAs (Protected Areas) considering the recommendations of Rodgers and Pawar (1988), BNHS' recommendations for IBAs (Important Bird Areas) and ATREE's study on KBAs (Key Biodiversity Areas);

Biodiversity Documentation

(1) Quantitative inventory of the flora and fauna, (2) detailed ecological study on the endemic and threatened species of plants and animals and their distribution maps, (3) formulation of conservation plans for the rare and endemic species.

Impacts of Development Projects

(a) River valley projects: 1) Evaluation of the existing hydroelectric projects in the Western Ghats; their impacts on the forest, biodiversity, hydrology and ecology, 2) cost-benefit analysis of the project, 3) study on the flora, fauna of the various river basins and along the course of the rivers till their confluence with the sea, 4) systematic evaluation of the down stream ecology and hydrology of the riverine systems, with priority to those rivers which are being considered for construction of hydroelectric dams, 5) status of the tanks, ponds and ιakes down stream

(b) Pesticides and the industrial effluents: 1) Monitoring the use of various pesticides and their effects on the biota, 2) demonstration of alternatives to chemical farming, 3) regular monitoring of the water bodies, including the reservoirs and those rivers running through PAs, 4) regular monitoring of the air quality, especially in the surroundings of industrial establishments,

(c) Development schemes: 1) Evaluation of the various schemes for tribal welfare, their economic benefits to and the impact on social fabric of the tribal and the general ecology of the area, 2) assessment of the impacts of the projects sponsored by various funding agencies such as World Bank, Asian Bank of Development and other external sources.

Ecosystem Structure and Function

(1) Studies on the structure and process of the various ecosystems of the Western Ghats, (2) long-term studies and monitoring of the various ecosystems, (3) life cycle and climate changes, (4) monitoring of ecological succession in abandoned plantations and estates, (5) ecological changes brought in by the exotic plants.

Socioeconomics

(1) Dependence of the tribal on the forest wealth and the "symbiotic relationship between the tribal and the forest/animal community", (2) socioeconomics of the forest dwellers - tribal- and the settlers, (3) ecologically sustainable livelihood, (4) sustainable harvesting of medicinal plants from the buffer zones of the PAs, (5) ecologic and economic assessment of the sustainable use of Non-Timber Produces.

Man-animal Conflicts and Eco-corridors

(1) Man - animal conflicts and sustainable solutions to them, (2) creation and maintenance and or strengthening of corridors between forests fragments.

Ecosystem Service Values

(1) Evaluation of the ecosystem service values of the various ecosystems of the Western Ghats, (2) case studies to demonstrate the economic loss due to the loss of forests, and other ecosystems,

Restoration of Damaged Ecosystems

Eco-restoration of the ecosystems damaged by anthropogenic activities.

Management of PA.s

Specific studies to provide essential information for management of individual PAs.

Awareness Programmes

(1) Initiation of a well laid programme for nature education for the school children, members of the Panchayat, Assembly and Parliament and officers at various levels and the general public, (2) strengthening of the Green Corps initiated in the schools by the Ministry of Environment and Forests.

Management Authority

Initiation of a Western Ghats Authority with representatives from the Government and Non Governmental Organizations from all the states covering the Western Ghats to coordinate the activities and liaise with the various stake holders.

These are some of the thoughts on the efforts required for the conservation of the Western Ghats. The workshop may consider them, delete or add wherever required.

SPECIES RECOVERY AND RESTORATION

Species recovery programs are viewed as an effective tool to restore critically endangered plant species. The program entails a) introduction of carefully genetically composed founding populations to the sites where the targeted species is likely to survive and/or, b) replenishing the existing populations with carefully genetically composed propagules. Such programs are hence characterized by the features of both ex situ and in situ conservation methods and also derive benefits of both. However effectiveness of these restoration programs is contingent upon our understanding of autecological details (such as the basic biology, nursery requirement etc), population genetic structure (allelic diversity, heterozygosity), demography (aspects of population size, sex ratio, population change etc.,) of the targeted species and construction of spatially explicit distribution maps of these species. It is also essential to carefully select the genotypes for reintroduction based on robust molecular data.

Ultimately a long-term persistence of the introduced population is important to the success of a recovery programme. Although a number of plant species of the Western Ghats are critically endangered, a very few isolated attempts are initiated towards the recovery of the endangered species. Majority of these attempts however, encompass only simple reintroduction of laboratory grown seedlings into habitats disregarding the genetic evaluation. Long-term implications of such introductions to the demography and genetic structure of the populations are not understood. Here we provide a general methodology that could be followed while recovering an endangered species. The steps would include: a) acquisition and build-up of the propagules b) identification of suitable niches for reintroduction, c) creation of genetically robust populations, d) monitoring the fate of the introduced individuals and e) up-scaling the programme. Threatened plants exhibit a myriad of life history traits hence this design should only be viewed as a general guideline and suitable modifications need to be incorporated on a case-by-case basis.

RECOVERY OF THREATENED PLANT SPECIES IN KOHLI AND SILENT VALLEY

Major ecological restoration will not be undertaken unless human society approves the goals and objectives of restoration. In addition, restoration will not persist unless human society has sufficient esteem for the restored ecosystem to protect its integrity. Landscape success is a measure of how restoration (or management, in general) has contributed to the ecological integrity of the region or landscape and to achievement of goals such as the maintenance of biodiversity. Attempt to derive ecological information on diversity from landscape level to species level is an onerous one requiring nesting of information prevalent across a hierarchical scheme. But the need to evolve a comprehensive approach to handle the variability of diversity across scales is growing due to the range of threats. Presently, anthropogenic land use change is contributing directly or indirectly to increased extinction risk for many countless plant and animal species. Geographic distributions of species and their relation to environmental variables will be a prime input for understanding the ecological niche of a given species. Association of particular species with particular environmental conditions has long been known, but quantitative analysis has been possible only recently, with the advent of tools for ecological niche modeling, as well as broad availability of remotely sensed information.

GARP (Genetic Algorithm for Rules set Prediction) or other ecological niche models can be used to identify regional limitations and managed habitat associations via the fine-grained landuse/ land-cover map and other bioclimatic variables. The environment uses GA / statistical model to arrive at different set of rules as alternatives with spatial inputs, for modeling the fundamental niche of a given species. Whereas field observations are obviously from a realized niche, the model points to the habitability of the individual, which should in turn be validated. Ecological niche modeling models of species' distributions visualized in geographic space are expected to provide a guide for future sampling.

Species distribution modeling of RET (Rare, Endangered and Threatened) species using DIVAGIS / DOMAIN / FLORAMAP/ GAM / GLM / GARP / Why Where can identify the probable areas of their occurrence for conservation purpose. There is an increase in environmental and ecological literacy is essential to achieving this end, as is an awareness of the goals and conditions of sustainability. If the science and technology underlying ecological restoration are not understood by the general public, implementation will fail for lack of public support.

RESTORING DEGRADED RAINFOREST FRAGMENTS

Historical fragmentation and a current annual deforestation rate of 1.2% in the Western Ghats biodiversity hotspot have resulted in a human-dominated landscape of plantations, agriculture, and developed areas, with embedded rainforest fragments that are important as animal corridors and biodiversity refuges. In the Valparai plateau, Anamalai hills, we selected within three rainforest fragments (5 ha, 19 ha, 100 ha) on private lands restoration plots ranging from open meadow and highly degraded areas (with dense Lantana camara invasion) to abandoned exotic tree plantations (Eucalyptus sp. and Maesopsis emenii) and plots with native and exotic tree canopy. From 2000, annual planting during the south-west monsoon was carried out of 7,538 nursery-raised seedlings in nine plots (0.25-1.5 ha).

The programme, a unique partnership with private landholding companies, also aims to encourage the use of native rainforest species as shade trees in coffee, tea, and cardamom plantations. Seedlings monitored at six-monthly intervals showed survival rates of between 34.2% and 90.7% under different site conditions over a two-year period. Seedling survival was higher in plots with complete weed removal as against partial removal along planting lines and higher in open meadow and under shade than in sites that earlier had dense weed invasion. Seedling mortality was higher over the dry season than the wet season in monitored plots. Retaining regenerating native species during weed clearing operations was crucial for rapid reestablishment of a first layer of canopy to shade-out weeds and enhance survival of typical shade-tolerant rainforest seedlings. Long-term sustainability of restoration on private lands requires incentives to participating landowners, identification of alternative fuel-wood resources for local people, and conservation education programs.

BIO-PHYSICAL CHARACTERISATION, CONSERVATION AND MANAGEMENT OF A RARE FOREST ECOSYSTEM

Myristica swamps are a unique fresh water ecosystem confined to low altitude, flatbottomed valleys drained by sluggish streams of Western Ghats The present study reveals the geographic distribution of individual Myristica swamps in southern Kerala, its mapping and its plant and animal diversity. A combination of survey methods using surveyors' compass, measuring tape and modern techniques such as GPS readings were used for delineating the swamps and fifteen swamps were accurately plotted in 1:50,000 scale Survey of India topo-sheets and area of each swamp was calculated.

About fifty three patches of Myristica swamps have so far been located from the study area. All the patches were below four hectares. For vegetation study, plots of size 0.1 hectare where tree species and GBH were recorded. Shrubs and herbs were enumerated from subplots.

A total of 65 tree species and 72 species of shrub -herb combination were recorded from the Myristica swamps. More than 60% of the trees belong to Myristicaceae family, the dominant species being Myristica fatua var. magnifica and Gymnacranthera canarica.

Other species of Myrsticacea and members of other families such as Celestraceae, Diptereocarpaceae, Anacardiaceae, Xanthophyllaceae, Myrtaceae, etc. were also present in the swamp, but less frequently. These swamps contain rare and endangered tree species such as Syzygium travancoricum, Myristica fatua var. magnifica. etc. Myristica swamp trees show adaptations such as knee roots and stilt roots. Ground vegetation belong to Araceae, Zingiberacea and Acanthaceae families. Reeds, canes and Pandanus are also present.

Various crabs, fishes, amphibians, reptiles and mammals found in the area were enumerated. In addition to this specimens of earth worms insects and mollusks were also recorded, though not exhaustively. Details of herpetofauna was looked into more detail. In spite of limited area distribution and tree diversity the Myristica swamps offer a variety of microhabitats for animals. In this study we analyse the variations in microhabitat preferences shown by selected herpeto faunal species of the Myristica swamps of Southern Kerala. Data collected during the first year of the study (March2004 - August 2005) has been used here. Random searches and transect sampling (100m x 4m transects) was done in the early morning and late evenings in the swamps. Forest patches adjacent to the swamps were also sampled to provide for comparison. All herpetofauna encountered inside transects were identified and the number per species and microhabitat were recorded.

Twenty three microhabitats in the swamp namely, leaf litter, grass, under log, bare ground, over boulder, water, under boulder, bark of tree, stagnant puddle, Pandanus leaf, Pandanus plant base, Gymnocranthera knee roots, Myristica stilt roots, tree trunk above 2 m, tree trunk below 2 m, wet sand, Lejenandra clumps, rotting logs, pebbled stream banks, marshy soil, exposed roots, herbs and lianas were classified inside the swamp.

Gymnocranthera knee roots, Myristica stilt roots and marshy soil were absent in adjacent forests hence twenty microhabitats were classified here. Percentage frequency of occurrence of each species per microhabitat was calculated. Niche breadth was calculated using Levin's formula.

It was observed that, in the Myristica swamps, species such as Nyctibatrachus aliciae, Nyctibatrachus major, Polypedates psuedocruciger, Rana temporalis and Trimeresurus malabaricus, which are generally considered as habitat specialists, were present in a variety of microhabitats. This was evident from high niche breadth values ranging from 18.09 to 8.64. Microhabitat preferences of these same species were also recorded in forest patches adjacent to the swamps.

Niche breadth values in forest patches adjacent to the swamps were significantly lower, the values ranging from 11.21 to 4.08. It may be concluded that the variety of microhabitats present in the Myristica swamps, provide these fragile animals more favourable conditions for survival and procreation, when compared to the adjacent areas.

The utilization of these microhabitats by selected herpetofauna, may have reduced intra- specific and interspecific competition in the Myristica swamps in spite of higher abundance when compared to the adjacent areas.

CONSERVATION OF ENDEMIC BIRDS IN THE SHOLA FORESTS

Loss of forest cover in the Western Ghats during last two decades was 25.6% while there has been a loss of 50% of shola forest since 1850. Western Ghats region has been recognized as one of the Endemic Bird Areas (EBA) with 16 species of endemic birds, including one endangered three vulnerable and four near-threatened species. Out of sixteen species of birds five species are restricted to montane shola forest, namely Nilgiri Laughingthrush, White-bellied Shortwing, Nilgiri Flycatcher, Black-and-orange Flycatcher, Grey breasted Laughingthrush while two species, Nilgiri Wood Pigeon and Crimson-backed sunbird are present in both shola and evergreen forest. Except the first one all others occur in the intensive study area Kukal.

It is one of the Important Bird Areas (IBAs) in the Western Ghats and it contains different habitats, namely shola (39%), grasslands (1%), wattle plantation (34%), pine (8%), eucalyptus (2%), agricultural fields (13 %) and others (3%). This study was carried out to understand the impact of plantation and fragmentation on the endemic birds. A 10 ha plot laid in shola of 80 ha area, 4ha shola fragment, five years old shola regeneration plot (5ha) and 10 ha monoculture exotic plantations (Wattle, eucalyptus and pine) were selected. Abundance of the endemic birds in each habitat was calculated by using circular plot method and habitat use was measured by recording foraging and breeding. Our study reveals that shola regeneration forest has more species diversity but abundance of the endemics was more in undisturbed shola. Except Crimson-backed Sunbird all the other endemics were breeding in the shola forest.

Usage of shola and the level of use were dependent on the size; small shola patches were used mainly for foraging and very rarely for breeding, whereas plantations were used mainly for resting and occasionally for foraging but not for nesting. Rregeneration plot supports more insectivores and nectarivores but not fruigovores, for breeding it supports shrub nesting species like laughingthrush. Increased human activities such as logging, fire wood collection and cattle-grazing affect the undergrowths of shola forest. Severe alterations of habitat adversely affect both feeding and breeding of endemics. Increased protection of the natural forest, shola forest, and restoration of such forest are recommended through community participation.

MARSHY GRASSLANDS IN THE WESTERN GHATS

Nilgiri Pipit is a rare and endemic bird species of the Western Ghats listed as near threatened (Birdlife International 2001). It is closely associated with short grasslands with marshy grounds mostly in the high altitude hill slopes (>1500m) of the Tamil Nadu and Kerala parts of the Western Ghats. Among 3321ha of surveyed area, 14ha of Upper Bhavani has the highest breeding density of the Nilgiri Pipit. Nilgiri Pipit breeds between March and June. It nests inside the marshy grasslands whereas the surrounding grass areas are used for feeding. The study revealed that this habitat is very important for nesting and other breeding activities of the species. These marshy areas are restricted to valleys of hills. Loss of the habitat to various anthropogenic activities is the primary reason for the decline in the population of the Nilgiri Pipit. The results showed significant preference of wet grasslands (marsh) and nearness to stream for the nesting habitat in the Nilgiri Pipit (chi-Square=27.867; $p<0.01$). 92.8% of the nests were in marshy grasslands in the study area.

Species diversity of grasses and herbs was high at the nest sites of the species. Andropogon sp. is the most used genus of the nesting plant for Nilgiri Pipit. Cattle grazing and there by destruction of its breeding habitat and nest were major disturbances in the intensive study area. Usage of nesting habitat by people for agricultural purposes and cutting the nesting grass for covering the tea are other major threats for the breeding pipits.

The available wetlands in the Korakuntha area and other areas have been reclaimed for agricultural purposes posing a threat to this species. Pipits used marshy grassland areas for all activities. Given this clear importance of high altitude grassland for Nilgiri Pipit, making management decisions and formulating land-use policies and there by conserving this endangered habitat have been recommended.

SURVIVAL OF TREE SAPLINGS IN DIFFERENT TYPES OF DEGRADED RAINFOREST RESTORATION SITES

Assessing the survival of planted seedlings is required to monitor the success of any rainforest restoration programme and to identify tree species that are suitable for restoration of sites differing in disturbance levels. In the Western Ghats biodiversity hotspot, we established in the year 2000 a programme to restore degraded rainforest fragments within private lands on the Valparai plateau, Anamalai hills. Restoration planting of around 95 rainforest tree species was carried out within three rainforest fragments of 5 ha, 19 ha, and 100 ha (7,538 saplings in nine plots of 0.15 to 1 ha area) following removal of invasive weeds. Here, we examine seedling survival of 46 rainforest tree species (>10 individuals of each planted) in fragments differing in levels of degradation, ranging from open meadow and highly degraded sites (with dense thickets of Lantana camara) to abandoned exotic tree plantations and sites with native and exotic tree canopy. Average survival after one year was excellent (>75%) to good (51-75%) in 33 (72%) and 10 (22%) of these species, respectively. Rainforest species seedling survival was higher over the monsoon than the dry season and higher in areas with better canopy closure. Regeneration in degraded sites with L. camara was dominated by the pioneer shrub Clerodendrum viscosum and lacked most tree species typical to undisturbed rainforests at that elevation. High survival of several typical mature forest species in abandoned exotic tree plantations and under shade of pioneers following weed removal indicates a high potential for relatively rapid re-establishment of mature-phase rainforest tree species using such restoration protocols.

DISTRIBUTION PATTERNS OF NEO-ENDEMIC DICOTYLEDONS IN THE WESTERN GHATS

The Western Ghats, also known as the Sahyadri Hills are well known for their rich and unique assemblage of flora and fauna. The Hill ranges have been recognized as one of the 25 biodiversity Hot-spots in the world. The senior author has undertaken a number of floristic studies in the Western Ghats,viz., Flora of Cannanore, Medicinal Plant Conservation Area Studies at Karian Shola, Topslip, Anamalais, Mathikettan Shola, Palani Hills, Mukurthi National Park, The Nilgiris are to quote a few.

These studies have helped in locating rich endemic areas, and also these studies could lead to the collection of a number of endemic, rare, threatened and endangered plants. In the present study, the authors have made an attempt to assess the taxonomic identity, distribution, niche choice, rarity, threat (if any), and they have a prioritized list of about 150 Neo-endemic dicotyledonous plants which are discussed in the paper.

ECOLOGICAL RESTORATION IN WESTERN GHATS

Ecological restoration is an intentional activity that initiates or accelerates the recovery of an ecosystem with respect to its health, integrity and sustainability. Frequently, the ecosystem that requires restoration has been degraded, damaged, transformed or entirely destroyed as the direct or indirect result of human activities. A properly planned restoration project attempts to fulfill clearly stated goals. Goals are attained by pursuing specific objectives of the restoration attempt.

To ascertain the fulfillment of goals and the objectives accomplished, evaluation or monitoring of the restoration effort has to be an integral part of the project. In the Western Ghats, restoration efforts are attempted by various agencies like governmental, non-governmental organizations and other self help groups. The goals of these projects are of varying degree based on their motives like restoring the biodiversity and restoring ecological and economic security for the benefit of local community.

However, most of these projects are not attempting to evaluate or monitor these projects. The restoration ecology would be better served in the region by recognizing that the diversity of conditions requiring restoration demands with much flexibility in goal setting, and that restorationists should seek to develop guidelines for monitoring these projects for the success of their efforts.

OPTIONS FOR MANAGEMENT OF WESTERN GHATS AS PROTECTED AREAS AND LIVELIHOOD ISSUES

A case for declaring the core western Ghat region of Karnataka as ecologically sensitive area (ESA) under the Environmental Protection Act of 1986 was found necessary considering the various ecological and socioeconomic characteristics.

The concept of ESA offers flexibility in conservation to allow the ongoing anthropogenic activities without causing damages to the unique ecology of the region. With the help of satellite data and physical survey of the region an area of 8000 km2 was demarked as ESA and detailed study of the ecology, geography, socioeconomics and other criteria for strengthening the argument for declaring the region as ESA was prepared. The results of the study show that the core Western Ghat region meets all the 13 criteria listed by the Ministry of Environments and Forests considering the proposal. The paper discusses some of the issues pertaining o selection of area, issues involved, attitude of the Government, and experience pertaining to areas already declared as ESA zones.

INVERTEBRATES AND CONSERVATION MANAGEMENT

Red List for Habitats and Green Lists for Species: A Top Down Strategy for Conserving Invertebrates

Although the concept of biodiversity shoot up to this popularity more than a decade back our conservation policies are still pivoting around the concept of conserving the game mammals. While considering the conservation, we forget grasses and sedges and invertebrates, which comprise more than 95% of the species. Many species invertebrates are being discovered from extremely small areas of Tropics. So the potential diversity of these lesser known taxa in the tropics will be very high and most of those are endemic to very small area. So if a relatively small area is logged or disturbed, many species will disappear forever. Since the majority of species remain undescribed and data on the distribution and abundance of only a very few described species are available, assessing the threat status of each species according IUCN norms is not practical. So for conservation, rather than species, habitats should be prioritised.

Butterflies of Shola Grasslands and their Conservation

The high altitude shola grasslands (>1600 ASL) of the Western Ghats are unique ecosystems with characteristic flora and fauna. Much of the current conservation effort in the shola grasslands focus on endemic mega fauna such as Nilgiri Tahr (Hemitragus hylocrius), laughing thrushes (Garrulax sp.) and other endemic birds. However, the shola grasslands harbours an unique butterfly community which require immediate conservation attention.

The endemic butterflies of shola grasslands such as Nilgiri Clouded Yellow (Colias nilagiriensis), Niligiri Grass Yellow (Eurema nilagiriensis), Palani Four Ring (Ypthima ypthimoides), Nilgiri Four Ring (Ypthima chenui), Red Disc Bush Brown(Mycalesis oculus), Red Eye Bush Brown (Mycalesis adolphei) and White Disc Hedge Blue (Calatoxia albidisca) are restricted to shola grasslands of southern Western Ghats. In addition to this, rare and non endemic butterflies of the Western Ghats such as Indian Fritillary (Argyreus hyperbius), Cabbage White (Pieris canidia) and Spot Puffin (Appias pandione) are also restricted to shola grasslands. Little information is available on the ecology these species.

Most of them are narrowly distributed within their range and depend upon a single larval host plant for their reproduction, making them vulnerable to habitat alterations. Habitat loss and deterioration of quality of remaining habitat due to expansion of plantation crops, introduction of exotic plantations such as Eucalyptus, Black Wattle and Pine, and fire in summer are immediate threats to the conservation of these butterflies. The paper discusses possible conservation and management options for shola grassland butterflies.

IDENTIFYING AREAS OF CONSERVATION IMPORTANCE IN WESTERN GHATS

Conservation in India has always been centered around charismatic mega- fauna approach. This has also been the case with the Western Ghats region which has sanctuaries dedicated for tigers, elephants, mountain goats to squirrels. It has also become a subject of status to very state within the country to dedicate a sanctuary to a large mammals particularly tigers. The untested assumption is that the large mammals by virtue of their large area and multiple habitat requirements will capture the diversity of other smaller organisms. Smaller organisms such as insects play enormous ecological role right from pollination, nutrient cycling to herbivory therefore it is also very vital to maintain their populations in all regions. However conservation of insects is also a biggest challenge as it involves a multiple species approach within each group such as ants, beetles etc. whose ecological requirements is difficult to fathom and is beyond one's scope.

However butterflies among insects have been an exception. By virtue of its charisma the minimum ecological requirements of butterflies have been recorded. Also their life history is strongly linked with their larval foodplants and most of them have specialized habitat and bioclimatic requirements. Western Ghats comprise 300 odd species of butterflies whose requirements if put together could be much more fined grained and capture more ecological niches than single large mammal species habitat.

Therefore butterflies could serve as a good probe taxa for other group of insects. In this paper we evaluate if the sanctuaries set aside or areas identified as conservation importance for the Charismatic mega-fauna will sufficiently capture butterflies in the Western Ghats region. We chose to evaluate butterflies against tiger, elephant and Lion Tailed macaque which represent varied trophic levels and life histories. Results indicate that the critical habitat of butterflies does not completely overlap with the areas, which have been earmarked as important for the 3 mega-fauna. Sanctuaries do not represent sufficient area of the critical habitat of the butterflies.

STATUS AND CONSERVATION OF FRESHWATER MACROINVERTEBRATES OF THE WESTERN GHATS

Aquatic macroinvertebrates play a significant role in the function of the wetland ecosystem. The freshwater bodies, especially the streams and rivers of the Western Ghats are recognized as global freshwater biodiversity hotspots for macroinvertebrates such as freshwater crabs and mollusks. However, very little information is available on the diversity, distribution and biology of most of the species. This has lead to under representation of this important group and their habitat in biodiversity conservation agenda of the region. Based on recently conducted studies on freshwater mollusks and stream insects along the entire stretch of the Western Ghats, the paper discusses the Integrating Science and Management of Biodiversity in the Western Ghats current status, threats, conservation and management options for freshwater biodiversity in general and freshwater macroinvertebartes in particular.

CONSERVATION OF THE SOUTHERN BIRDWING BUTTERFLY TROIDES MINOS CRAMER IN THE PEECHI VAZHANI WILDLIFE SANCTUARY, KERALA INDIA

The Southern bird wing butterfly, Troides minos Cramer is endemic to the Western Ghats of South India. Out of a total of thirty-six of bird wing species, eleven are listed in threatened categories and five including Troides minos Cramer as requiring further research and monitoring. The various aspects of the biology and ecology of this species which were relevant to its long-term conservation were evaluated in this study. The biology of the species was studied by rearing out the different stages in the laboratory. The larval feeding preferences and mortality factors in each instar were noted. Descriptions of the different larval instars along with details of the dimensions of the head capsules were compiled.

In order to obtain information on the specific requirements of Troides minos Cramer with respect to the other closely related Troidines occurring in the same habitat, the guild approach was adopted in ecological monitoring. Data on the abundance, seasonality and the effect of weather parameters on sightings of species of the Troidine guild viz., Pachliopta aristolochiae Linnaeus, P. hector Fabricius, P. pandiyana Moore and Troides minos Cramer over a three year period between June 1998 and May 2001 were obtained. Analysis of the relationship between sightings of the Troidines and temperature, humidity and rainfall was done to understand the weather parameters governing habitat suitability for these species.

The population of Troides minos Cramer in the Peechi and Vazhani areas was monitored and the mean monthly counts and indices of abundance obtained. Males of Troides minos Cramer were found to aggressively defend resource patches and behavioural sampling was done during the period of peak aggressive encounters in the territories during 1999 and 2000. The various behaviours were categorised and represented as kinematic diagrams and analysis indicated a resource-based territoriality in this species.

The larval feeding preferences were assessed by studying larval weight, larval period and pupil weight of juveniles reared on three hosts viz., Aristolochia indica Lin., A. tagala Cham. and Thottea siliquosa (Lam.) Ding Hou. The significance of nectar resource in the establishment and occupancy of the species within territories was evaluated by correlating the number of inflorescences of Clerodendrum paniculatum Lin. and occurrence of chasing and feeding behaviours. Conservational measures which included assessing the effect of the introduction of alternative larval and adult host plants as well as the captive rearing and breeding of the species are discussed.

13

Scelionidae of India-An Overview

Scelionids form an important group under parasitic Hymenoptera. In the role of natural enemies, as egg parasitoids, they check the population of a wide range of insects belonging to orders: Lepidoptera, Orthoptera, Diptera, Coleoptera, Heteroptera and like, whose members mostly qualify as pests of considerable economic importance. Of the three subfamilies viz., Telenominae, Scelioninae and Teleasinae, the first two outnumber the third, both in diversity and abundance. Among Telenomines, Telenomus Haliday, with an array of hosts is the most commonly encountered group, in Scelioninae, Gryon Haliday, Scelio Latreille and Baryconus Foerster are much frequented, whereas Teleasinae is mostly represented by Trimorus Foerster. Other than commenting on the general techniques employed in scelionid collection, along with some interesting attributes of the family, this paper aims to compile the hitherto studies on Scelionidae of India and discusses the current status of the taxon in India and at the global level.

EFFECTS OF RAINFOREST FRAGMENTATION ON SPIDERS

Studies looking at the problem of tropical forest fragmentation and disturbance have tended to focus on vertebrate groups particularly birds and mammals, and plants. There has been a paucity of research on invertebrates in this regard despite the fact that they make up the major portion of the diversity particularly in tropical rainforests. Spiders are one such group of invertebrate predators in tropical ecosystems that are known to be sensitive indicators of environmental change. The present study makes a preliminary assessment of spider species diversity and community structure and its response to rainforest loss and fragmentation in the southern Western Ghats, India.

Seven rainforest fragments in private land undergoing various stages of degradation, three relatively undisturbed rainforest fragments falling within the Indira Gandhi Wildlife Sanctuary and two coffee sites were sampled by visual searches along a belt transect during the dry and pre-monsoon period between January and May 2005. A total of 4565 individual spiders (4300 detections) were recorded within a sampled area of 5.757 hectares. These comprised of 156 identified morphospecies belonging to 21 families and 8 designated functional groups. There was little difference between the sites in all the measures of spider diversity used - species and individual density, rarefaction and estimated (Jackknife 1) species richness. However, cluster analysis of Bray-Curtis similarities in spider species composition of sites revealed four distinct clusters - high altitude undisturbed, low altitude undisturbed, low altitude disturbed and coffee sites. Species which contributed significantly to the dissimilarity between undisturbed and disturbed forest, and forest and coffee sites were identified which could possibly be used as indicators of habitat alteration.

INSECT CONSERVATION THROUGH PUBLIC PARTICIPATION

GREEN CLUB is a Non - Governmental, voluntary, organization dedicated for the conservation of nature. Green club has link with many Educational institutes, universities and other NGOs to reach its goals. As individual and as species, the insects outnumber all other groups of animals combined. Over a million species of insects have already been named and described.

Greenclub conducts studies on insect diversity at Bandipur National Park. The survey was conducted in 4 phases in the month of October, December, April and June. During this survey insect specimens were collected using traps like, Yellow pan Trap, Pitfall Trap, Food lure Trap, Pheromone Trap, Light Trap, Malaise trap, Insects were also collected by use of insect nets and many were collected through handpicking and hand brushing method. Collected specimens were preserved identified by use of standard taxonomic keys and with the help of experts. So far we have completed species level identification of ants, dung beetles, butterflies. The information generated through this study is disseminated to public through slide shows, exhibitions and other interactive programs.

DIVERSITY AND ABUNDANCE OF SELECTED TEPHRITIDAE ALONG KERALA WITH SPECIAL REFERENCE TO WESTERN GHATS

Systematic and quantitative surveys are generally lacking in India for any group of organisms. The present study is perhaps, one such rare effort to cover the Western Ghats Integrating Science and Management of Biodiversity in the Western Ghats area of the southern most State, Kerala to survey for a group of insects, members of the tribe Dacini coming under Tephritidae (Diptera). The survey was conducted using the two parapheromones, methyl eugenol and cue lure, during two seasons along sixty-seven locations of Kerala.

The survey yielded 1738 specimens of fruit flies coming under 12 species of three subgenera viz, Bactrocera Macquart, Zeugodacus Hendel and Javadacus Hardy under a single genus Bactrocera Macquart. Of the twelve species seven collected at methyl eugenol traps belonged to a single subgenus Bactrocera and five at cue lure traps were distributed among all the three subgenera encountered. The diversity and richness of the species encountered during the course of survey have been mapped using GIS software. The diversity of fruit flies measured both in terms of species richness and the Shannon's index indicated significant negative association with the latitude. The unusual trend might be a reflection of need for greater and intensive sampling in relatively small geographic areas. Further, the two parameters were also found to show lack of association with the altitudes.However, the geographic niche breadths of individual species was strongly associated with the diversity of their recorded host plants, which explains why some species are more widespread in their distribution.

Arthropod Communities as Monitoring Tool for Restoration Success

Ecological restoration enjoys widespread use as a technique to mitigate for environmental damage. Success of a restoration project often is evaluated on the basis of plant succession only. Recovery of the native arthropod fauna is an important sign that the effort achieved restoration goals. Assessing the progress of restoration projects with arthropods has many advantages. Arthropods have been recognized as efficient indicators of ecosystem function and recommended for use in conservation planning. Assessing restoration projects presents a special challenge for the use of arthropods as bioindicators, because such projects usually involve planting some component of native vegetation, certain taxonomic groups may respond quickly to such efforts. The restoration ecology would be better served in the Western Ghats region by seeking to develop guidelines for monitoring these projects for the success of their efforts.

IMPACT OF ANTHROPOGENIC DISTURBANCES ON LAND SNAIL DIVERSITY OF WET EVERGREEN FORESTS

Disturbances are common feature of forests of developing tropical countries such as India. Primary forests such as evergreen forests of the Western Ghats, have undergone a major transformation due to several anthropogenic activities. Land snails have a major role to play in soil nutrient recycling. Here we have assessed the impact of anthropogenic disturbance on land snails of the Western Ghats. We sampled 88 plots classified into less disturbed and highly disturbed plots based on composite computed threat index. Less disturbed sites are more diverse and have high abundance of snails compared to highly disturbed sites. There is significant difference between diversity, richness and abundance between two sites. Five species dominate the land snail community in both the disturbance categories, and contribute 62.26 per cent and 72.30 per cent of the total individuals in less disturbed and highly disturbed sites respectively. Majority of the species occur in low densities. There is a high degree of dissimilarity between less disturbed and highly disturbed sites. Plots within highly disturbed sites are more variable in terms of species composition. Twenty nine percent of the species recorded were micro gastropods and there is a significant difference in the abundance of micro and macro gastropods between less disturbed and highly disturbed sites. Snail richness is negatively correlated with composite threat index.

INTEGRATED APPROACHES TO MINIMISE AMBIGUITIES IN ANURAN DESCRIPTION

The Western Ghats and Sri Lanka, among the hottest hotspots of biodiversity, are being in more focus on amphibian studies, especially after the report of discovery of about 200 old world tree frogs from these two regions. Eventually, these regions became amphibian hotspots of the globe, surpassing Borneo, Madagascar etc.

The evidence of decline in amphibian global population fuelled amphibian research in the past two decades, which has led to the description of new species and also to controversies associated with the approaches in describing a new species. Review of the genus Philautus in 2001, provided a comprehensive account on the genus, which has heralded a competition among researchers from Sri Lanka and Western Ghats' on Philautus diversity. Amphibian diversity studies were also supplemented with the new discoveries of caecilians and other ranids in the Western Ghats.

Such taxonomic oriented research activities in these regions, though is a positive development from conservation point of view, but has led to controversies due to ambiguous ecological status of some species. This has serious consequences in the conservation and management of Western Ghats, which necessitates an immediate step to arrive at a protocol in describing a new species. This paper while documenting the integration of many disciplines and emphasises the need for arriving at a consensus before assigning 'new status' to a species and ascertaining its ecological status.

In the Western Ghats, since 2000, 18 new species of amphibians belonging to two extant orders, namely anura and gymnophiona have been described mainly on the basis of morphology and morphometry (Mor). Recent advances in technology have provided sophistication in analysis and interpretation of molecular data (Mol), acoustics/vocalization (Voc) and osteology (Ost) to describe a new species. Recent description from Sri Lanka and Western Ghats indicates fragmented approaches in describing a species even today despite availability of better methods. However, most of the species have been described incompletely as there are limitation in traditional as well as in newer methods. This necessitates integrated approaches involving morphological, molecular and acoustical analysis along with natural history and osteology in species description.

The case of so called 'extinct' frog taxa Philautus leucorhinus, Lichtenstein, Weinland and von Martens 1856 from Sri Lanka highlights the need for integrating molecular and morphological approaches, supplemented by osteological, acoustical and developmental analysis in species description. In this case, except morphometry, all relative information like vocalization, reproductive behaviour and molecular data are either obscure or not available, creating unnecessary confusion among conservation biologists on the exact status of the species.

Apart from the mandatory requirement of vouchering a type specimen to national museum with the information on place, date of collection, collector name and ecological notes, it is essential to standardise the protocol integrating all the aforesaid methods with keys to assign 'new taxon' while describing a species. This would eliminate the lacunae associated with the traditional as well as advanced methods. Morphological and molecular will complement each, while other methods like acoustics, natural history and osteology would supplement the description, resulting in a comprehensive description of the species. This would certainly eliminate the controversies associated in describing anuran species.

ECOSYSTEM SERVICES

The dispersal of seeds that ruminants regurgitate in dry tropical forests In dry tropical forests of the Indian subcontinent ruminants disperse over 10% of the woody flora. Mostly small seeds (< 5 mm) have been found in a viable state in the ruminant dung, while large well-protected seeds are reported to be regurgitated intact by ruminants. Due to a constraint on size of particles that can pass out of the ruminant forestomach, large seeds have to be either ground down to a passable size or disgorged during rumination. While the role of ruminants in serendipitous dispersal of smallseeded grass and herbaceous plants has received some attention, ruminant-mediated dispersal of large seeds is a scarcely studied process in tropical forests.

Ruminant-mediated dispersal of large seeds was examined via Phyllanthus emblica-Axis axis interaction. A captive deer regurgitated intact seeds of P. emblica after retaining them in the rumen for 7 - 27 h. A considerable fraction (22%) of deer-regurgitated P. emblica seeds germinated, although this was lower than that of unconsumed seeds (72%). The size and resistance to cracking of seeds like P. emblica might ensure that ruminants regurgitate them intact instead of defecating them, thus ensuring dispersal over long distances. Ruminants, among the few frugivores known for many large-seeded plants, are more liable to be affected by habitat degradation than smaller frugivores. Thus, local extinctions of ruminants can leave plants like P. emblica without effective long-range dispersers.

Insect Pollination Services in Natural and Semi-natural Landscapes is Affected by Pollinator Type and Inter-plant Distance

Ecosystem services sustain life but rapid and widespread changes in the world as a consequence of human activity tend to overburden and undermine the natural capacity of ecosystem members to provide these services. Despite their tremendous value and vulnerability to landscape changes, we have little detailed ecological understanding of pollination services. Using case studies from Bhimashankar Wildlife Sanctuary, a highly fragmented seasonal cloud forest in the Western Ghats of Maharashtra, we examine how wild pollinators respond to changes in land use and habitat alteration and thereby influence fruit set of plants in natural and semi-natural landscapes. In semi-natural landscapes the habitat has been altered considerably by thinning natural vegetation for crop cultivation, fuel wood and timber needs of the local human population.

We examine pollinator visitation rates and fruit set in three trees species with different breeding systems that are insect-pollinated. Diospyros montana (Ebenaceae) is a dioecious tree that is pollinated by a host of flies, beetles and bees. Lasiosiphon eriocephalus (Thymeleaceae) is a self-compatible, bisexual shrub pollinated principally by nitidulid beetles and Heterophragma quadriloculare (Bignoniaceae) is a self-incompatible, bisexual tree pollinated by carpenter bees. Results from the three species indicate that although plant breeding system is an important determinant of pollination success in altered landscapes, the scale at which pollinator services are affected by landscape changes vary for different pollinator taxa and depends on the mobility of the pollinators in relation to inter-plant distances.

BAT POLLINATION IN THE SOUTHERN WESTERN GHATS

Bats act as pollinators as well as seed dispersers in the southern Western Ghats. So far six species of phytophagous bats have been identified in the southern Western Ghats. These are Latidens salimalii, Eonycteris spelaea, Rousettus leschenaulti, Cynopterus sphinx, C. brachyotis and Pteropus giganteus. Plants adapted to bat pollination through nocturnal anthesis, emit strong odours to attract bats, secrete copious nectar for consumption and produce large amounts of pollen for bat consumption and crosspollination. The Indian short-nosed fruit bat C. sphinx is found in the plains and up to 600 m elevation in the southern Western Ghats. In Ceiba pentandra trees, bat activity is maximum in the early part of the night due to the availability of the nectar, which in general diminishes in the later part of the night. Cynopterus sphinx usually hovers and lands on the inflorescence and laps nectar from the flowers.

The head and the ventral side of the bat are covered with pollen during this process to facilitate cross-pollination. Studies on the pollination of C. pentandra revealed that bat pollination is 40 % more efficient than self-pollination and 25 % more efficient than insect pollination. In Bassia latifolia, the bats remove the fleshy petals with attached stamens and fly to the nearby night roost to feed on them. The pistils remain intact on the plant along with the other floral parts. The muzzle of the bat is dusted with pollen in this process, thus facilitating cross-pollination. Foraging peaks on this tree species were observed only in the first part of the night. Although foraging by C. sphinx in B. latifolia appears to be destructive, it is effective in pollination, since pollination by bats yields 20% more fruit set than selfpollination. The conservation of bats not only facilitates forest growth through seed dispersal but also give provides benefits through pollination of economically important plants.

Variation in Trophic Cascade with Distance from Natural Forests and Importance of Higher Trophic Levels in Ecosystem Services: Natural Pest Control

Severe anthropogenic perturbations have upset several balanced states in ecosystems leading to issues such as pest outbreaks. In a tritrophic system, predators and parasites are crucial in maintaining the population size of primary consumers to an optimal level. However, a pest outbreak might occur with even slight reductions in the richness and abundance of the predators and parasites. Very few studies have evaluated the services rendered by higher trophic levels in controlling pests. This study is an attempt to fill this lacuna. Here we address the importance of predators and parasites in the context of ecosystem services by studying trophic cascades that include wood-boring curculionid beetles as primary consumers and their predators and parasites. We studied the effects of predator species richness and density on the intensity of attack by borers on four species of trees, viz. Grewia tiliaefolia, Grevillea robusta,

Xylia xylocarpa and Tectona grandis in relation to the proximity of these trees to natural forests stands. We show that the predator species richness and density increased with proximity to natural forest stands. Moreover, there was a considerable increase in borer infestation with decreasing predator species richness and density.

LESSER KNOWN ENGINEERS OF TROPICAL FOREST ECOSYSTEMS

Frugivory and seed dispersal are key ecological processes in tropical ecosystems. Fruits are a major food resource for a variety of taxa, and are predominantly dispersed by vertebrates. Considerable research has been conducted on the role of birds and primates as seed dispersers and their impact on tropical plant communities. However, recent studies have indicated that other vertebrate groups also play a significant role as seed dispersers in tropical forests. One of these lesser known groups is the viverrids, particularly the members of the sub-family Paradoxurinae or palm civets. During an ecological study of a nocturnal viverrid species endemic to the Western Ghats, i.e. the brown palm civet Paradoxurus jerdoni, in the rainforests of Kalakad Mundanthurai Tiger Reserve, it was observed that the species was predominantly frugivorous. Fruits formed over 90% of the civet's diet throughout the year. These civets do not damage ingested seeds that are dispersed within their home ranges (average of 20 ha) in the rainforest habitat. Seed germination experiments on a subset of the civet's dietary species indicated enhanced germination success and germination rates in some cases. Here, I discuss the role of such viverrids as an integral member of the frugivore community in tropical rainforests by assessing their effectiveness based on the breadth of their diet and in comparison with other known avian and mammalian frugivores such as hornbills and macaques.

Can Pollinators alone help good Fruit Setting in Teak?

Studies conducted in KFRI on the role of pollinators in the pollination ecology of teak revealed that they are inevitable for better fruit setting and for viable seed formation. Teak prefers cross pollination and the rate of fruit setting varied depending on the number of pollinators involved in the pollination process. There was slight variation in the faunal diversity of insect pollinators in pure teak plantations and natural forest with teak. The study on the pollination ecology of teak could identify a variety of insects involved in the pollination of teak, insects damaging teak flower buds, premature fruits and a number of other insects visiting teak inflorescence. The study revealed that in the absence of sufficient pollinators the fruit production would be very less. Incidence of fungal pathogens was also found to cause reduction in fruit setting. Fungal infection by Alternaria sp and Phomopsis sp. was noticed. Depending upon the duration of infection and location of the plantation the fungal infection rate varied between 10 to 90 percent. The infection attained its climax during the monsoon season. As a result of fungal infection seed setting was poor often leading to the premature fruits fall. In the fruit-cutting test it was noticed that less than 30 percent of fruits were found to have viable seeds. Fortnightly application of selected fungicides on the inflorescence from the early flowering to fruit maturation stage was found to give good control of the fungal disease.

What Pollinates Embelia Ribes and how Successful is It?

Embelia ribes is a polygamo-dioecious species with prolific flowering. Studies on reproduction were conducted in about sixty individuals in the Silent Valley National Park, Kerala. Male flowers have a highly reduced pistillode and well developed anthers. Female flowers are with pistils as well as anthers that contain no pollen (cryptic dioecy). During early mornings, pollinators in larger number visit male plants first and then the females owing to difference in time and the amount of nectar presented. Wasps are the most common visitors followed by social and wild solitary bees. The species exhibits a moderate fruit to flower and seed to fruit ratio. Flower abortion ranges from 60-86%. Pollinator limitation is a major cause for flower abortions; pollen-pistil interaction studies reveal absence pollen on stigmatic surfaces. Fruit set ranges from 14-40%. Fruit abortiveness ranges from 1-27%. The species exhibits a male biased sex ratio

Avenues of Research on Mycorrhizal Conservation in the Western Ghats

Nowadays, Biodiversity Conservation has gained centre stage because of the worldwide destruction and disappearance of natural resources. It is evident with the dwindling of population of some endemic species like Hornbills, Lion-tailed Macaque and many higher plants including Medicinal and Ornamental plants. Our think-tank and policy making are highly inclined towards such macro-level organisms, which are relatively at the tertiary levels of ecosystem. This paper addresses the need for more research on mycorrhizal conservation in the Western Ghats.

Although, ecosystem functioning involves high level of dynamic inter-relationships, it is the lower organisms including microorganisms and insects that make a significant contribution for its stability. When it comes to the management of Natural ecosystem, it is the fungi that facilitates and channels the nutrient cycling within the system. The fungi have many important functions, largely unrecognized, in the biosphere. Many species are beneficial saprophytically and symbiotically (Mycorrhiza and Lichens). They degrade the lignin and cellulose in plant residues and initiate the decomposition of these complex compounds, thereby enabling nutrient supply to plants. Field surveys have found that plants (about 50% of Angiosperms, 25% of Gymnosperms and a few Pteridophytes) with mycorrhizal associations predominate in most natural ecosystems throughout the world. Most information about the mycorrhizal status of plants is from the Northern Hemisphere in temperate regions, especially for crop plants and weeds.

The mycorrhizae of tropical regions including Western Ghats are less studied. Although about 13,000 species of mycorrhizae have been recorded in this region, its role in nutrient cycling is not yet completely studied. Epigeous and hypogeous sporocarps of mycorrihzal fungi are important food sources for mammals. Mycorrhizal roots and fungus fruit bodies are important as food sources and habitats for invertebrates. Mycorrhizal fungus hyphae are an important food source for soil invertebrates.

Since most mycorrhizae are sporulators, weather conditions and nutrient availability in the form of litter mostly control their physiological activity at the community level. By understanding mycorrhizal species composition it is possible for tracing their habitat requirements. Western Ghats forests, already being gene banks of mycorrhizal fungi needs more attention on basic research on above aspects for Forestry and Agricultural applications, thereby maintaining the status of Western Ghats as a World Biodiversity hotspot. This paper elaborates on the possible avenues of research, which is not yet been touched till now which includes the modern Biotechnological and Bioinformatic applications.

WATER AND FORESTS

The Need for Further Hydrological Research in the Wet, Mountainous Sahyadri Ranges

The hydrology of the Sahyadri (Western Ghats) regions in Karnataka, and in parts of Kerala, seem to be far different from that mountains of very heavy rainfall and mountainous areas elsewhere in the tropics. The Sahyadri areas show much slower catchment response and much less surface flow, than even in those the areas in extra tropical regions with comparable rainfall magnitude. This fact is evident from the values of Drainage Densities (below 3) and the shape of the streamflow hydrographs, which indicate 'baseflow' contributions exceeding 60%. Such very strange features of the hydrological status of the region may be attributed to (i) the intensity of rainfall being low, despite the quantity being very high, a feature non-typical of the tropics, (ii) highly porous and very deep soils, very rarely found in mountainous regions and (iii) dense vegetative cover on appalachian slopes and in the valleys and a good cover of grass on slopes in the escarpments.

The runoff processes in the region are best explained by the Jones' Extended Variable Source Area Theory, according to which while flow through pipe-nets below the surface contribute significantly to streamflow, surface runoff is mainly generated by rainfall on contributing areas riparian to the stream. Investigations by this author have shown that surface runoff generated due to outflow from pipes away from the stream channel leads to additional infiltration excess surface runoff, a process usually not recognized elsewhere.

Lessons learnt from the findings including the above are (i) that methods of hydrological analysis and watershed management practices presently being used in the area need to be revised thoroughly and (ii) that further investigations can help understand better some important processes, which are unique to this region in the tropics. Unfortunately, it seems very little is being done.

Impact of Iron ore Mining on Suspended Sediment in a Tropical Catchment in Kudremukh, Western Ghats

This study utilized secondary and primary rainfall, stream flow and suspended sediment data to investigate the impacts of iron-ore mining within the Kudremukh National Park on the Bhadra river in south India. Available secondary suspended sediment data from the 1980s was analyzed together with more current data collected during 2001-2003 at instrumented sites upstream and downstream of open-cast mining activities (~4.2 km^2). More than 50% of the suspended sediment load in both the Bhadra River and Bhadra Reservoir comes from mining-affected lands that occupy < 1% of the total catchment area of 1968 km2. For baseflow conditions during the post-monsoon period of 2001, 0.02 Mg km-2 day-1 of suspended sediment was discharged at the upstream site compared with 0.74 Mg km-2 day-1 downstream of the mine. During the 2002 and 2003 monsoons, these rates increased to about 1.99 and 7.89 Mg km-2 day-1 for upstream and downstream sites, respectively. The specific sediment yield above the mine is 239 Mg km-2 yr-1 versus 947 Mg km-2 yr-1 downstream. Sediment concentration downstream was significantly higher than upstream for all conditions. The current annual suspended sediment load below the mine ranges from 100,000 to over 150,000 Mg, depending on the size and frequency of large rain events.

During the 2002 and 2003 monsoon seasons, 20-30% of the total suspended sediment load during the sampling period of 67 and 123 days respectively was transported in one single day and over one-third of the total recorded suspended load is discharged in less than 4 days in each monsoon. Daily rainfall and maximum hourly rainfall intensity were reasonable predictors of daily sediment loads downstream of the mine (R2=0.71 and 0.575, respectively; $p<0.001$). However, the upstream response was not predicted well by these variables, suggesting an absence of rapid overland pathways and/or reduced availability of easily transportable sediment on the less-disturbed lands above the mining area. Large episodic sediment events downstream of the mine were associated with daily rainfall > 150 mm and hourly intensities exceeding 20 mm. One hundred and nine such episodic events occurred between 1990 and 2001 alone. A conservative estimate of the total suspended sediment load in the Bhadra River after mining commenced in the 1980's is 1.37_106 Mg, although the actual value could be considerably higher, up to 107 Mg. Comparison of historic data and another study in 1994, with recent measurements confirm that mining and associated activities in Kudremukh National Park are the greatest sources of sediment entering the Bhadra River. The Bhadra river carries considerably more sediment now than before mining started, damaging riverine ecosystems and disrupting downstream water resources. Note: Findings from this study are currently in press.

Baseflow Analysis

The major problem in the management of rivers has been how to balance the tradeoffs between instream (e.g. auqatic life, and recreation) and out-of stream (e.g, reservoir regulation) uses. Management problems normally exacerbate during low-flow periods and with on-going water resources development resulting in gradual reduction of flow available for instream uses.

There are number of studies carried out to analyse the possible environmental effects caused by instream flow reduction. Such reduction may lead to increased sedimentation that changes the morphology of the stream channel and flood plain. Changes in stream morphology may potentially affect the distribution and abundance of stream biota. Streamflow reduction can also aggravate the effects of water pollution. The components of a flow regime which are considered important for the estimation of Instream flow requirements include low flows, small increase in flow and small and medium floods. This paper describes a simple technique for baseflow separation from continuous daily stream flow records. The technique employ the smoothed minima algorithm, which has been widely for baseflow separation. Also, it describes the possible application of baseflow in the context of water resources management.

In the present study is an attempt has been made to relate the baseflow index with nondimensional topographic parameters derived for 17 catchments in Western Ghat regions of Karnataka state, with areas ranging from 5 km^2 to 600 km^2. The result shows that, the catchmnet size has no effect on baseflow index. However, the regional regression model, indicates that, the total stream length, length of main stream and slope index dominates the relationship between baseflow index.

Hydrology in the Western Ghats: Why is it Important and what do we Know and don't Know?

The Western Ghats region of India is one of the twenty-five global hotspots of biodiversity. These mountains are the sites of the heaviest rainfall in peninsular India- the streamflow from which generates most of the hydropower and provides lifesupporting water to most of the 245 million people that live in the peninsular Indian states. The region contains more than 30% of all plant, bird, and mammal species found in India. These forest ecosystems also generate significant benefits to local communities. These forest and grassland ecosystems have undergone extensive conversion to nonforest land-cover and the remnant areas are subject to intensive human-use with major negative impacts on biodiversity and ecosystem services. The biodiversity values of the Western Ghats ecosystems have received local, regional, national and international attention. Almost all hydrologic components such as streamflow, ground-water recharge and evapotranspiration are of critical concern to human livelihoods, ecosystem functioning and biodiversity.

However, the hydrologic services of the different natural and human-modified ecosystems are poorly known.

A review of all peer-reviewed research published in the past few decades suggests that we know very little about the basic hydrological processes and effects of land-cover change, watershed services and socio-economic impacts at various spatial scales for the Western Ghats region. However the published work has important contributions to the understanding of both hydrological processes and small catchment response. These include evapotranspiration of individual trees, effects of land-use/land-cover change on river and reservoir sedimentation, sub-surface flow through natural soil pipes, and the impacts of converting montane grassland ecosystems into monoculture eucalyptus plantations.

The major gaps in our knowledge include estimates of evapotranspiration of different forest and human modified ecosystems and how this is changed by forest degradation, degree of deciduousness and land-use, the impact of retaining or losing certain types of forest and grassland ecosystems on sedimentation of reservoirs and the effects of reduced infiltration due to grazing and other land-uses on ground-water recharge and evapotranspiration. Other important gaps are the variability in rainfall intensity in relation to soil hydrologic properties and the linkage to catchment hydrologic response. A concerted effort needs to be made to encourage relevant hydrologic research and basic data collection in the Western Ghats as the first step in informing policy and decision makers so that the political and civil society resolve to conserve forests and grasslands in the Western Ghats is enhanced.

14

Impacts of Forest Cover Loss on Hydrologic Processes

This study focuses on investigating the hydrological impacts of land-uses (primarily grazing and fuel wood extraction) within Bandipur National Park in Karnataka, India. These land-uses are proximate causes leading to loss of forest cover and replacement by degraded scrub forest, and as such, representative of much of the Indian forested landscape. The resulting forest cover loss has substantial impact on the radiation energy balance, the water balance and partitioning, and on soils. The impacts are relatively undocumented in India, despite the potentially tremendous implications on local and downstream water resources. The deciduous nature of these original and replacement forest covers adds another significant challenge to understanding the concurrent hydrological changes-especially of evapotranspiration.

The focus of this recently initiated study is to (1) investigate the differences in soilphysical and soil-hydraulic properties that relate to hydrologic processes of infiltration and surface runoff; and (2) monitor evapotranspiration, in degraded and protected forest areas within the Park. The research will complement ongoing rainfall and streamflow monitoring in the same region. Research results will aid in forest and water management decision making, and the local population in understanding the ramifications of their practices on their water resources.

My presentation will cover the background on the hypothesized impacts of forest cover loss on soil and evapotranspiration, the research design, and the challenges in understanding the ramifications at a larger regional scale.

ROLE OF LAND SURFACE PROCESSES ON HEAVY PRECIPITATION

We examine the effects of land surface processes on precipitation including heavy rain episodes. We first present an overview of the theoretical bases of land surface processes that affect regional precipitation. We then present findings, which related to land surface and land use/land cover change, have impacts on micro, meso and continental scales. Results from two recent case studies are highlighted. The first case considered is the impact of urbanization on thunderstorms and heavy precipitation episodes in the vicinity urbanrural interface. We show observational (radar, satellite, gage data) evidence of the impact on precipitation structure and distribution. We then discuss results from several numerical experiments to further explain the impact of urbanization on convection and precipitation. The second case we consider is the June 26, 2005 heavy precipitation case over Mumbai, India. Results from several ongoing numerical experiments suggest that land surface processes have significant impacts - perhaps even larger than topography - on the heavy rain event. We conclude the presentation by providing additional perspectives regarding future configurations for better representation of land surface processes and their associated dynamical feedbacks. The discussion will conclude with examples regarding configurations with aerosols and radiative feedbacks, and consideration of satellite datasets that can improve the model performance.

JOINT FOREST PLANNING AND MANAGEMENT IN WESTERN GHATS REGION

In the Western Ghats region, 226 Villge Forest Committes were sampled, accounting for 80% and 41% of VFCs formed in the district and Western Ghats project respectively to study the institutional issues. The study encompassed addressing issues related to spread of JFPM in the region, formation and functioning of VFCs, involvement of communities in protection and management of forests and impact of JFPM on vegetation and social and economic impact on the community. The perspective of communities with regard to impact of JFPM was also assessed.

There was conceivable increase in the density of vegetation due to JFPM in the Western ghats region. The communities also perceived an increase in fuelwood though there was no discernable change in NTFPs and considerable decrease in grass availability. There was marked improvement in the relationship of community with the Forest Department.

POLICY IMPERATIVES FOR THE WESTERN GHATS

Settled courses of action of social groups or governments adopted or followed for expediency or advantage in specific areas of activity are called policies. Policy statements are articulations of the approach or an expression of the attitude of societies, polities or governments to be consistently followed for a long period. They are all framed for material benefit and stated to be for the common good.

We have in our country over the past century framed many policies regarding resources or collective properties such as water, forest etc vested on the State for management on behalf of the people. Such policies are essentially for guidance in consistent, continuous management. Laws, regulations and administrative procedures have to reflect these policies. Evolution of such policies is an important element of the dialogue essential in democratic decision making in civil societies.

Policies regarding natural resources are mostly restricted to economically valuable resources or exploitable components of our environment. Hence they all suffer from large lacunae. These lacunae may be with respect to the incompleteness of the theme covered or due to the inadequacy of the understanding of the issue dealt with. At times policies suffer from the ad hoc approach intended for shortsighted problem solving. In our country most of our existing policies regarding environmental components, both physical and biological, suffer from ecological inadequacy. Continuous natural processes are disarticulated and treated as independent events. Complex integrated systems are dealt with focussing on individual components. Natural system boundaries are not taken into consideration; instead man made temporary administrative or political boundaries are treated as definitive frames of reference for management.

These flaws are partly due to 'convenience' in decision-making. At times they are due to lack of scientific understanding. In most instances policies reflect the common understanding of an earlier period. Policies can rarely be futuristic or even realistic. Because they are often framed to tackle a crisis or in response to en event, they fail to pre-empt an emerging problematic situation. Hence they fail to help the society interface with a constantly changing scenario and attain an acceptable dynamic equilibrium. But this has to be our aim in framing policies.

In our country we still do not have clear cut policies on landuse to regulate extensive landuse changes in response to market motives. We do not have policies to protect soil or the quality of environment, rural or urban. We are far from enshrining in policies concepts such as Wilderness or the integrity of natural landform. All the abovementioned examples have serious repercussions in areas such as the Western Ghats.

Stepping into another dimension, we have to remember that we are all short-term inhabitants of a small planet with limited resources and ever expanding needs. We also cannot afford to forget that our capacity to inflict damage on the biosphere is expanding even faster. Policies have to transgress national boundaries and tackle biospheric dimensions. Generational environmental justice to most of us is still a very utopian concept.

The biospheric unity demands beyond limited regional perspectives and national sovereignty global visions, unprecedented strong decisions and radical shifts in human mindsets. But we also know that our track record, even during the recent past, even when catastrophic happenings such as global climate change has become a reality, is not exactly confidence instilling.

The Western Ghats

The Western Ghats is perhaps the most diagnostic component in the landscape of peninsular India. It exerts its all pervasive influence on every other terrestrial landscape unit in this part of the country and its influence even extends on to littoral marinescape. Weather and climate, regional hydrological cycle, living systems, human communities in this tract and all our developmental activities impinging upon millions of people are strongly influenced by this hill chain. The Western Ghats has distinctive characteristics and retaining this fundamental identity is essential to sustain all its positive influences on the environmental processes modulated by it. Its role as the sub-continental watershed divide is primary. From Godavari in the north to Tambraparni in the south, all the peninsular Indian east flowing rivers originate from the eastern face of the Ghats.

Similarly the very large number short torrential streams disgorging from the western face of the Ghats flowing into the Lakshadweep Sea are under the control of the Western Ghats. Interception of the monsoon stream and precipitation of rainfall is the most fundamental environmental influence of the Western Ghats on the ecology of this part of the country. Hence any policy regarding water for this part of the country has to take cognisance of this reality and regulate human activities to prevent undermining this role. Western Ghats because of its very topography, climate, altitudinal and latitudinal zones, biogeographic and human historical antecedence has an unusual assemblage of ecosystems, habitats and species of biota. The geographic location and the evolutionary history of the Indian landscape have gifted the Western Ghats with a biodiversity notable for its high endemism. Terrainal constraints have helped retain even now, even though fragmented, scattered and under pressure, benchmark ecosystems particularly tropical rainforests in the southern parts.

Over the past decades, particularly since the 1970s, popular environmental concerns have strengthened. But simultaneously the threats also have multiplied. Hence it is all the more reason for an in-depth introspection on the state of affairs of the

Western Ghats and review existing measures including policy directives to safeguard its natural bounty. The most vulnerable and valuable potential of the Western Ghats is its natural forest cover. The remaining primary rainforest and associated ecosystems particularly in the southern reaches of the Ghats need special attention in the regional as well as international biodiversity conservation action. Developmental projects, particularly those designed to make use of the water potential of this hill chain have inflicted maximum damage on its natural ecology. This trend still continues and policies regarding river impoundment and river diversions need to be critically reviewed and corrected.

Even from the larger Indian context, forest cover of the Western Ghats has been of great importance since the colonial period. Forest management focussing on timber resource exploitation has been directing all our forestry activities for the past couple of centuries. This has inflicted terrible damage on the ecology of the Western Ghats. Even after realising the unsustainability of the current forestry practices we continue to deal with the forested landscape of the Western Ghats merely as a timber mine. Even though mentioned in passing in the current Forest Policy, the radically different approach to forest management in areas like Western Ghats, which has to become ecosystem management, has not happened. In fact, due to the past mismanagement, practically all the remaining forested tracts in the Western Ghats require long-term complex restorative management. Although community participation has been stated often enough it is yet to become a corner stone of the hill area land husbandry.

The Western Ghats because of its biodiversity potential also has the largest network of Protected Areas in Peninsular India. Yet even in the grossly inadequate extent of forest land brought under such legal protection, both policy and management of biodiversity conservation remain grossly inadequate and often enough counter productive. Even under the relatively better situations it remains large mammal management and catering to tourism needs. Sustaining biodiversity is nowhere in the picture. In the name of what has been called ecotourism, justifying it all by the unproven claim of income generation, biodiversity protection and forest conservation are slipping into a dangerous quicksand. The rapid rate of urbanisation taking place in the Western Ghats in recent times is yet another set of human activity which needs regulation through comprehensive policy change.

Initially it was hill stations such as Ootacamund and plantation district headquarters such as Munnar which stood out as urban centres in the otherwise pristine landscape. In the Post-Independence period a number of Project Townships in association with river valley projects sprung up all over the interior reaches of the Ghats. Not only they expanded with uncontrolled influx of squatter population, whole new illegal urban development took place as in the case of Idukky district in Kerala. Mining centres and even religious centres started expanding in various stretches of the Ghats. Kudremukh and Sabarimala are examples. Of late recreational hill resort development has reached epidemic levels in various parts of the Western Ghats. The destruction of Ooty and its environs and the devastation around Sabarimala have not taught us any lessons. We are yet to have a comprehensive, ecologically sound urban development policy in position in the Western Ghats. Not only that all the existing urban centres have gathered an uncontrolled runaway momentum, additional ones are being planned by the government itself with no safeguards and at times even without any justification.

In association with urban development or otherwise, large-scale construction activity is accelerating in the Western Ghats and inflicting terrible damage on the terrain, ecology and biodiversity. Road construction is the most widespread construction activity taking place now both through governmental involvement and on public initiative. In most instances there is no EIA and in many cases the activity is patently illegal. Along with road construction, quarrying of rock and other construction material such as laterite is also expanding everywhere in the Ghats. There need to be the formulation of strong policy guidelines regarding construction activity in the Ghats.

The Western Ghats is one of the most important locations in India as the habitat of a large number of tribal communities. Although there are policies for tribal development, rehabilitation of displaced people etc, taking into consideration the specific characteristics of the Western Ghats, locally relevant modifications of existing policies regarding tribal development is urgently needed. The currently debated Scheduled Tribes (Recognition of Forest Rights) Bill 2005 is a specific example.

Policies are statements of the intention of the society. Even when there is a clear-cut policy statement, practices could be the opposite, destructive. Similarly even if there is no policy statement, a society can consistently do the right thing unprompted by law. So all that really matters in the long term is the understanding and vision of the human community. In all matters regarding environmental protection the world is facing such a terrible crisis not merely because of the paucity of laws, policies or scientific knowledge, but because of the total confusion in the collective human mindset regarding where we should be heading for. What is called as developmental crisis is actually an inner human conflict, our inability to resolve the conflict as to what we truly need to take out of this earth and what we can take out just because we have the means to take it out. When many more of us have decided that a little less is more than adequate, the problem resolves on its own. Till then we need to have policies and laws to check the erosion of the foundation.

CONSERVATION AND LIVELIHOODS

The Western Ghats presents a clear case of declining forest wealth due to increasing developmental activities in the region. The present study based on a socioeconomic survey of the villages surrounding the core Western-Ghat region of the Karnataka state covering 8000 km2 spread over in five districts shows the growth of commercial activities and increasing population density. Over the years the GDP from forest, value of timber, non-timber forest products and NTFPs have declined drastically. Thus, the decline in the overall NTFP collections and decline in livestock population indicates the ecological vulnerability of these products to commercial harvesting practices.

The declining trend also reflects the socioeconomic vulnerability of the local communities dependent on such resources. The households within the core region enjoy better water supply and sanitation facilities. Further the there has been a faster growth in the urban infrastructure over the years. The study observed that key changes that affect the sustainable development of the region are Irreversible land development, decline of environmental conditions required for renewable products, diminution of drinking water quality due to rapid urbanization, socioeconomic instability of traditional communities and finally decline of profitability of eco friendly production.

Some Aspects of Breeding Ecology of the Common Myna

The breeding ecology of Common Myna (Acridotheres tristis) in and around Mayiladuthurai was observed between January 2005 and June 2005. The morphometry of Common Myna nests viz., length, width, depth and the weight were observed. The percent availability of nesting materials were indicated that herbs (34.18 %) were the predominant nesting materials followed by fibres, roots, twigs, papers, feathers, leaves and others. Totally 43 nesting-site were observed and the height of nest-site, nest height and DBH were measured individually in each nest. The availability of different habitats such as agricultural lands, water sources, bare land, groves, human habitations and road and distance to electric post independently influenced the nesting-site selection. The clutch size, length, width, weight of the eggs, incubation period, hatching success and fledging success were measured in each nest. Various nestling morphometric measurements such as weight, body length, bill length, bill depth, wing length, wing span, tail length, tarsus length and middle finger length were measured thrice in a week. The weight of the nestlings increased across the day and attained the maximum weight of 106.38 g on day 29 then it gradually declined to 99.2 g on day 33. The other morphometric measurements showed a gradual increase across the days.

COFFEE BASED AGROFORESTRY SYSTEMS

Kodagu, a beautiful land along the Western Ghats is the second smallest and one of the densely wooded districts in the country. The commercial timber harvesting from the natural forests and introduction of coffee cultivation was initiated by Britishers during nineteenth centaury. The traditional methods for the establishment and maintenance of coffee plantations involve placement of young coffee plants under canopy provided by one or more tree species and diverse coffee plantations can be considered as managed forests. In the last decade due to increase in the coffee price, the area under coffee cultivation was also increased at the cost of natural forested ecosystems. Some people believe that these coffee based agroforestry systems are looming threat to our forested ecosystems. On the other hand, it is also believed that, they harbor trees, so help in conservation of biodiversity. In the present study the answer for the question to what extent they help in conservation is experimented and also been revived from the earlier studies from Kodagu.

CONSERVATION EDUCATION

In 'Biophilia' Prof. E.O. Wilson, brings out our propensity to affiliate with other life forms. This innate tendency of biophilia (i.e. the love of humans towards other life forms) is the most effective force for nature conservation. 'Nature Education' by definition aims to inculcate the love for nature in the human mind. The best way to conserve the nature is to facilitate the love towards it. Love towards nature develops when we interact and involve in the activities in such a way that nature reveals the wonders it holds. The two streams of such activities are that of science and art. Though humans are biophilic in nature, the present day world is more tempted to drift away from it.

From view of our own survival and long-term welfare, we need to conserve nature and the best method is to facilitate our own biophilic tendencies. Last five years of experiences of SACON's Nature Education Programme provides a way for it. From the experience with the children and public of Coimbatore, confidently the author confirms the biophilic nature of people.

Moreover, the tendencies are stronger in the village than the cities. We need a different set of methods and tools for the nurturing of biophilia. The difference here is that of the approach and the style of execution. The paper presents a few case studies to highlight the differences. The author presents the best approach is to attempt involvement the students and public in the process of learning from nature through scientific methods, artistic expressions, interactions and even conservation action.

CEE's Initiatives for the Conservation of Western Ghats

Centre for Environment Education (CEE) a national Institute established in 1984 as a Centre of Excellence in Environmental Education, supported by the Ministry of Environment & Forests, Government of India and affiliated to the Nehru Foundation for Development is engaged in developing programmes and materials to increase awareness about the environment. CEE through its various activities is mainly focussed in the following areas.

Solid Waste Management

CEE- Coorg initiated a solid waste management programme to address the garbagerelated problems of the towns of Kodagu which has resulted in the degradation of soil and were having the impact on the flora and fauna of the region. CEE has set-up a poly bag weaving and waste paper recycling unit at Coorg District to demonstrate the concept of reusing the waste materials. Similar initiative has been taken up by CEE- Kannur in Kannur district of Kerala. Solid waste management programme initiated by CEE Central in Pune Municipal Corporation, with support from SKF India Ltd and as part of Waste Matters, a consortium of NGOs in Pune.

Interpretation

CEE's interpretation programmes enhance the nature experience by providing on-site information through signages and exhibits.CEE- has developed a Kodagu Heritage Interpretation Programme (KHIP) in Madikeri for interpreting the natural as well as the cultural uniqueness of Kodagu district of Karnataka. CEE-Tamil Nadu Sate cell has developed an interpretation centre for Indira Gandhi National Park and Wild Life Sanctuary at Aliyar which showcases the biodiversity, the threats it is facing and cultural heritage of the indigenous pecple of the region. CEE-TNSO has also set up wayside signages from Aaliyar to Valparai depicting the flora and fauna of the region.

Documentation

CEE took up a project called Sacred Groves of Kodagu: Role of students in Conservation and Documentation, in 160 primary ar ' higher secondary schools in three taluks of Kodagu, to document the biodiversity of the groves and to educate school children about their importance.

School Education

CEE has undertaken many projects such as 'Anandshala', in Gujarat, School EE programmes in Goa along with many partner agencies including government departments.

Camping Activities

CEE-TNSO has organised a series of Nature Education Camps for school and college students. These camps were meant to engage students in positive outdoor experiences Integrating Science and Management of Biodiversity in the Western Ghats that would develop an attitude among students for caring and conserving environment amongst students.

Organic Farming

CEE- Kannur has taken up organic farming initiatives in Kannur under which it promotes the use of liquid manuring, vermicomposting and other biodynamic methods to increase crop yield.

On the Tools of Communicating Nature

Conservation education reaches success when it accomplish in greening the hearts of people. For reaching the young hearts, conservation educators use various elements of nature as tools. In this paper, the author attempts to highlight relative positive and negative effects of various elements of nature as tools of communicating Nature. The author finds that the birds and butterflies are the better tools to percolate in to the heart of children.

MNHS Activities for the Conservation of Nature

Malabar Natural History Society (MNHS) is a not-for-profit, non-governmental, voluntary organisation dedicated to Nature education, research and conservation of biodiversity. It is based at Calicut and is very active since its inception three years back. As part of its Nature Education and Research Programme, it has published three books. It also publishes a quarterly newsletter; Malabar Trogon, regularly conduct trekking and thematic Nature camps. To create environmental awareness it also organise slide talks for students, teachers and common people.

Its members include all kinds of people from students to professors, taxi drivers to software engineers, masons to scientists; all united with one common interest, Nature. The book Butterflies of Kerala, published by MNHS in the regional language was the first of its kind. It was a trendsetter in the nature education and conservation arena of Kerala, and turned many students, teachers and public to ardent butterfly watchers and nature lovers. With the help of its enthusiastic members, MNHS has successfully established the migration route of butterflies in the Western Ghats for the time. MNHS has also launched a 'Common Bird Monitoring Programme (CBMP) to study the biology and ecology of the common birds seen around us and but neglected even by the professional ornithologists.

BIODIVERSITY MONITORING THROUGH COLLEGES

A case history of a participatory approach in biodiversity conservation through the college students presented here. A study of inventorying and monitoring of biodiversity of western ghats was initiated in 1994 by Prof Madhav Gadgil of Centre for Ecological Sciences in Indian Institution of Science,

Bangalore. As the part of this programme, authors have initiated a study with college students in one panchayat at the foothills of Ernakulam District of Kerala. The study conducted with systematic protocols. The fiveyear study resulted in documenting the ecological history of the area and identified 863 species of plants, 21 mammals, 202 birds, 18 reptiles, 16 amphibians, 56 fishes and 124 butterfly species. This study has become an eye opener to the local people as it revealed the biotic richness of the area. Local people are now associating with the forest department and have initiated a programme for the sustainable utilization of the biological resources in the area. Plans are on the anvil to develop a wilderness learning programme and an environmental education project by the panchayat showcasing the biodiversity. This can be sustained further by the cooperation of the academic and research communities.

THE ACTIVITIES OF A NATURE CLUB AND THE CHANGES IN ATTITUDE OF STUDENTS

This paper presents the story that how the exposure to nature revolutionized the attitude of students in Kadri mills Hr. Sec. School, Coimbatore. The Salim Ali Nature Club introduced in the school in the year 2002. The author describe how dry was the earlier experience of teaching biology and the changes that happened after the series of field trips and student field studies carried out by the children of their school. The paper also high light the various approaches in the activities carried out in the school nature club and effect of these approaches on students. Along with the scientific activities, the effect of the nature club activities involving arts and literature in reaching out the imagination of the children is also emphasized.

Communicating Nature through Arts

In this paper, authors demonstrate how the arts and literature can be used as the effective medium for communicating the sensitivity that is required for the nature conservation. The nature camps of artist students exposed the hidden but colourful world of nature through the structure and life of birds and butterflies to the students. The students enjoyed the camps and later continues to reflect nature in their all the later works and life. The other students those who see the excellent art works of their friends also gets motivated and attracted to wonderful and colourful nature of birds and butterflies. The children exposed to nature's wonders express more effectively through the medium of literature also. The posture is presented to highlight the caliber of students to communicate Nature in the most effective way.

SPECIAL SESSION ON FOREST MANAGEMENT

Though India has inherited a rich legacy of conservation, and could boast of a unique approach adapted by our ancestors of living in harmony with nature as opposite to the human desire to conquer the elements of nature, it could be said that the year 1864 heralds the era of scientific management of forests in India. On this year the forest department was created by the British rulers of India, and Dietrich Brandis became the first Inspector General of Forests. The first Forest Policy was laid down by Government of India in 1894, which had a definite bias for serving agricultural interests. Fortunately, ever since, the policies and principles of management have been subjected frequent to modification in an attempt to recognize the needs of the hour.

One of the major shifts in forestry education is the appreciation of the complexity of the web of life in the forest ecosystems. It needs be noted that the interdependence of life forms and the intricate ways in which diverse organisms are connected to one another was appreciated by the modern world very recently and that Ecology is a new branch of science. Even dictionaries published as late as 1960s do not contain such terms as "environment" or "ecology". India possesses a distinct identity, not only because of its geography, history and culture, but also because of the great diversity of its natural ecosystems. It is only in recent times that the forestry training has given due emphasis towards appreciation of the rich biodiversity at the molecular, species, ecosystem and landscape levels of the country. It should be heartening to note that the curricula of the training institutions of forestry now give due emphasis on understanding the megadiversity of the country.

Our planet is suffering from a variety of environmental ills, with issues like global warming, ozone layer depletion, toxic chemical emissions, acid rain, erosion, land degradation, air, soil and water pollution and a variety of such issues. In fact, the planet earth is now heading towards a mass extinction of species, India being no exception to this global phenomenon. Models of species extinctions (and predictions of loss of species in numbers) lead to some ominous predictions. When applied to tropical areas (forests in particular), the models suggest that in the next 25 years or so, 20-25% of the various groups of organisms (especially plants and birds) will be wiped out. In this backdrop, yet another shift in the training of forestry personnel is to sensitize them towards species extinction, the causes that push extinction processes and grapple with this problem by imbibing in them the knowledge and skills of ex-situ and in-situ conservation. The conventional approach of forestry with a focus on revenue has changed to the conservation of natural resources, and the focus on production has long ago shifted to sustainability focus. There have been paradigm shifts also concerning decision making processes, management practices, and development of human resources and this paper discusses as to how the forestry courses at various levels are moving towards a focus on holistic management of our forests and biodiversity.

MANAGEMENT OF BIODIVERSITY IN THE WESTERNGHATS : CHALLENGES

Western Ghats is one of the areas of endemism and rich biodiversity in our country. The Nilgiris District is having the largest geographical area under forest land use management. The network of hydroelectric projects in the area, depend on the catchment protection and water harvesting potential of the forest areas.

Plantations of Eucalyptus species; wattle and tropical pines taken up in the Nilgiris from the last century has been a source of debate as to their boon or bane. The Tamil Nadu Forest department has stopped plantations of exotics from early nineties. Naturalisation of the original vegetation through phased removal of the exotics is being followed. Ecological restoration of the grasslands in the Nilgiris plateau has been studied and the restorative process has been documented.

The role of fire in the ecological sustenance of the dry deciduous forests in the Mudumalai and segur regions is also an important management perspective. The role of livestock grazing in the Masinagudi area and the conflict with Wild Life management is also getting due attention through various studies. The effect of rampant damage to the rich forest areas of the Gudalur Wyauad has not been focused so far. An integrated holistic approach to conserve the various land use management strategies in the first Biosphere reserve of the Country is the need of the hour.

The research wing of the TNFD was established in the year1918. Experts in forestry management had handled forestry research in state, and the research wing had met the emerging forestry challenges from the time to time offering the best solutions for field adoption. In the earlier phase of the research, forest resource survey, research on Eucalyptus, Wattles, and other fast growing species were taken up to meet the needs of people and the growing needs in the state. Later, research was carried out in areas of NTFP, sandal,teak and other economically important timber. Introduction trials and evaluation trials across various agroclimatic zones were taken up to asses the quality of timber introduced.In the third phase, clonal forestry was given importance.

Simultaneously, studies on population dynamics in various forest types were undertaken to study the forest patterns and ecological changes from time to time. Presently the thrust of research is on lesser known timber species and secondary timber species to promote agro forestry. The other areas of research include organic tree husbandry, tree improvement post harvest technology and restoration forestry.

The research wing of the forest department has a definite vision plan and forest research policy to guide research activity in the state, the broad areas of research being biodiversity conservation and bio-productivity enhancement. In the Western Ghats extensive studies have been taken up by the research wing to study the Shola Grassland association and changes in floristic composition over time due to human interference. Collaborative research covering various areas of concern in the Western Ghats have also been taken up.

The gene pool garden in Gudlur, Nilgris is an in- situ, cum ex-situ research center for the RET plant species of western ghats.Research on medicinal plants have been taken up in many research centers like Brook Hampton,Glen Morgan in Nilgris and Palakulum in Kanyakumari.

The nursery technology and seed biology for various species of endangered sholas have been under taken by the research wing. Conversion of monoculture Eucalyptus and pine plantations to polyculture through under planting have shown good results.Studies also been taken up for recovery of RET species, fire control strategies and soil micro organisims for developing suitable and appropriate technologies for restoring degraded sites in the Western Ghats.

The research wing is supported by the network of forestry extensions centers in the state. More areas of the research have been planned for floristic and ecosystem development in the Western Ghats for the future, which are discussed in this paper.

The Malnad region of Karnataka State is part of the Western Ghats, an area famous for its biodiversity with a rich endemic assemblage of plants, arthropods, reptiles, amphibians, birds and mammals, many of which remain unknown to science. The forest is also the home of the largest venomous snake in the world - the much respected and feared King Cobra.

Much of the king cobra's habitat has unfortunately already been lost to deforestation is response to increasing demand for timber and agricultural land, fuelled by spiralling human population growth. The remaining forests of the Western Ghats are heavily fragmented, and it is widely held that the area is in urgent need of research, conservation and education activity, both for the survival of the king cobra's rainforest habitat, and the welfare of the people dependent on the forest for resources and water.

The author has recently acquired eight acres of land surrounded by the last great tracts of rainforest in the region, upon which we are setting up a unique rainforest research station with the support of the Whitley Fund for Nature. This research and education facility will make long term field studies on all aspects of rainforest ecology, previously logistically difficult, a reality. Using the charismatic king cobra as a flagship species in the same way as the tiger is used elsewhere, the station will provide a logistics base for a wide range of forest studies, conservation action, local participation and education. The new station will not only enable scientists and students to maintain a scientific presence in the area, but will also become a place through which to network with the local people, NGOs and Government agencies involved in conservation of the region. It is anticipated that this model will be replicated in other parts of the Western Ghats and efforts in this direction have begun.

Biodiversity of the Western Ghats recorded so far:

Birds - 508 species, 4% endemic

Mammals - 127 species, 17% endemic

Reptiles - 176 species, 62% endemic

Amphibians - 219 species, 78% endemic

Fish - 218 species, 53% endemic

Arthropods - 338 butterflies & thousands more insects

Over 200 spiders

Trees & plants - 8400 species, 45% endemic

In the light of increasing anthropogenic pressures on the forests, protected areas, including national parks, sanctuaries, and biosphere reserves serve as last footholds for conserving biological diversity. However, because protected areas are often targeted for the conservation of selected species, particularly charismatic animals, concerns have been raised on their effectiveness in conserving non-target taxa and their genetic resources. In this paper, we evaluate if protected areas could serve as refugia for genetic resources of economically important species, which are threatened due to extraction pressures.

We examine the population structure and genetic diversity of an economically important rattan, Calamus thwaitesii, in the core, buffer and peripheral regions of three protected areas in the central Western Ghats, South India. Our results indicate that in all the three protected areas, the core and buffer regions maintain a better population structure as well as a higher level of genetic diversity compared to the peripheral regions of the protected area. Thus, despite the escalating pressures of extraction, the protected areas are effective in conserving the genetic resources of rattan. These results underline the importance of protected areas in conservation of non-target species and emphasize the need to further strengthen the protected area network to offer refugia for economically important plant species.

15

Pollinators: Neglected Biodiversity of Importance to Food and Agriculture

INTRODUCTION

Agricultural biodiversity is often understood as crop genetic resources, yet agroecosystems hold a wide diversity of other organisms that contribute toward their productivity and sustainability. Amongst these are pollinators, being animals that carry pollen from the male to the female parts of plants and thus ensure that fruit or seeds are formed. Over the past decade, the international community has increasingly recognised the importance of pollinators as an element of agricultural diversity supporting human livelihoods. Yet mounting evidence points to a potentially serious decline in populations of pollinators.

Maintaining and increasing yields in horticultural crops, seeds and pastures through better conservation and management of pollinators is critically important to health, nutrition, food security and better farm incomes for poor farmers.

This information document "Pollinators: Neglected Biodiversity of Importance to Food and Agriculture" has been prepared by the Seed and Plant Genetic Resources Service. The document presents the role of pollinator diversity in healthy ecosystem functions; its value for crop production, seed production, and forage resources and its role in adaptation to changing environments and minimizing risks for farmers.

Pollination and Ecosystem Functions

In nature, the vast majority of flowering plant species only produce seeds if animal pollinators move pollen from the anthers to the stigmas of their flowers. Without this service, many interconnected species and processes functioning within an ecosystem would collapse. With well over 200,000 flowering plant species dependent on pollination from over 100,000 other species, pollination is critical to the overall maintenance of biodiversity. Approximately 80 percent of all flowering plant species are specialized for pollination by animals, mostly insects. Greater dependence on animal pollination in the tropics and mountainous regions

The dependence of ecosystems on animal pollinators is even stronger in the tropics than the global average: less than 3% of all tropical lowland plants rely on wind for pollination. In the tropical forests of Central America, insects may be responsible for 95 percent of the pollination of canopy trees, and vertebrates (bats and a diversity of other taxa) may pollinate 20 to 25 percent of the subcanopy and understory plants, and insects a further 50 percent. Arid and mountain ecosystems often have highly diverse pollinator communities as well, with finely tuned adaptations to ensure that pollination is effective even when climatic conditions are erratic.

Not only are tropical areas of the world more dependent on animal pollinators, but they may also be more susceptible to pollinator loss. An international working group1 has shown that plants are especially likely to suffer lower pollination and reduced reproductive success in areas with high plant diversity, presumably due to an intense competition for pollinators in these diverse ecosystems. These include South American and South-East Asian forests and the rich fynbos shrublands of South Africa.

Diversity of Pollinators

The diversity of pollinators and pollination systems is striking. Most of the 25,000 to 30,000 species of bees (Hymenoptera: Apidae) are effective pollinators, and together with moths, flies, wasps, beetles and butterflies, make up the majority of pollinating species. Vertebrate pollinators include bats, non-flying mammals (several species of monkey, rodents, lemur, tree squirrels, olingo and kinkajou) and birds (hummingbirds, sunbirds, honeycreepers and some parrot species). Current understanding of the pollination process shows that, while interesting specialized relationships exist between plants and their pollinators, healthy pollination services are best ensured by an abundance and diversity of pollinators.

Pollination as a Crop-related Genetic Resource

Contribution of Pollination to Food Security

In agro-ecosystems, pollinators are essential for orchard, horticultural and forage production, as well as the production of seed for many root and fibre crops. Pollinators such as bees, birds and bats affect 35 percent of the world's crop production, increasing outputs of 87 of the leading food crops worldwide, plus many plant-derived medicines in our pharmacies. Role of pollination in expanding horticultural production

Food security, food diversity, human nutrition and food prices all rely strongly on animal pollinators. This is particularly the case of horticultural crops. Diversification into horticultural crops is becoming an avenue to poverty alleviation amongst many farmers around the world. The trade in horticultural crops accounts for over 20% of developing countries' agricultural exports, more than double that of cereal crops. Unlike the historical increase in cereal production, the expansion of production in fruits and vegetables has come primarily from increases in the area cropped, not from yield increases. The consequences of pollinator declines are likely to impact the production and costs of vitamin-rich crops like fruits and vegetables, leading to increasingly unbalanced diets and health problems. Thus, maintaining and increasing yields in horticultural crops under agricultural development is critically important to health, nutrition, food security and better farm incomes for poor farmers.

Pollination services make important contributions to other aspects of crop production. Improvements in quality of both fruit and fibre crops, such as cotton, are the result of good pollination. Deliberate management of pollination contributes to yields of oil for biofuels from new and alternate source (castor oil, and croton in Brazil, for example). Pollination in chilli peppers will contribute to an increased speed of ripening that can translate into getting peppers to market at a higher off-season price, and for one additional flush of fruit over the course of a growing season.

Growing Awareness of the Importance of Pollination

In the past, pollination has been provided by nature at no explicit cost to human communities. As farm fields have become larger, and the use of agricultural chemicals has increased, mounting evidence points to a potentially serious decline in populations of pollinators under agricultural development. The domesticated honeybee, Apis mellifera (and its several Asian relatives) have been utilized to provide managed pollination systems, but for many crops, honeybees are either not effective or are suboptimal pollinators. Managed honeybee populations are also facing increasing threats of pests, disease, and reluctance among younger generations to learn the skills of beekeeping. The process of securing effective pollinators to "service" agricultural fields is proving difficult to engineer, and there is a renewed interest in helping nature provide pollination services through practices that support wild pollinators.

Conventional wisdom has held that crops such as tomatoes and coffee are self-pollinated, and growers need not concern themselves with insect visitors. But when crops are grown under increasingly industrialized conditions, such as in greenhouses for tomatoes, or high-input sun coffee, the contribution that animal pollination can make to yield--or conversely, the losses when native pollinators can no longer reach the crops--become more evident.

Economic Estimates of the Value of Pollination Services

Economic calculations of the value of pollination services (commercial plus wild) vary widely but range from $112-200 billion annually at the global scale. In the United States alone, the annual contribution of wild pollination services alone has recently been estimated at over $3 billion annually5. Local-scale studies estimating the contribution of pollinators to crop yields found that:

(1) In Costa Rica, the enhanced value of coffee production attributed to wild bee pollinators living in nearby forests was estimated as $393/ha of forest, and this value was both an order of magnitude above existing government programs to pay for environmental services, and equivalent to other competing land uses such as timber production.

(2) In canola seed production in northern Canada, fields near uncultivated areas produce greater yields due to greater pollination services from a more diverse and abundant wild bee community. They calculated that farmers could maximize profits by retiring up to 30% of the field area from production, in order to receive higher yields on the remaining 70%.

In a time of decreasing commodity prices due to oversupply, yield gains may not always be the most important consideration for agricultural producers. Pollination should be a concern of farmers from the standpoint of quality, not just quantity. For pyrethrum, derived from the flowers of Chrysanthemum cinerariifolium, a more potent insecticide is produced when the flower heads have been visited by insects. In many countries, quality is vitally important, because well-shaped fruit fetches much higher prices in the selective export market. If such quality considerations can enter into market share and market prices, pollination may contribute significantly to the income per unit area for farmers conserving pollinator services.

The Ecosystem Context

Increasingly, it is realised that not just genetic resources alone, but the interactions between them create healthy agroecosystems. Pollination knowledge is distinctly ecological knowledge, and needs to be placed in an ecosystem context to be properly understood; it is neither solely about plant reproduction or insect visitation patterns, but rather about the interrelations. The interlinkages, while extremely important, make knowledge of pollination complex, and more like a network or information system than discrete bodies of knowledge. Often the most critical interactions that determine reproductive success of plants are often not the most obvious ones, and actions taken to conserve plants do not necessarily conserve their pollinators. Therefore, an ecosystem approach is needed, and information dissemination on pollination services should reflect an ecosystem context. Thus pollinator conservation entails promoting the awareness that not just species, but also the interactions between species merit conservation and careful management, as a way to strengthen key ecosystem linkages. Pollinator conservation underlines the importance of linkages between conservation of ecosystem functions, sustainable production systems, and poverty reduction.

Because restoration is far more difficult than conservation of existing interactions, a strong argument can be made in favour of conserving wild and indigenous pollination services in other systems before they are similarly lost. Management of wild pollination services requires an ecosystem approach with boundaries of the system drawn beyond fields, into the broader agroecosystem. Defining management concerns beyond the field limits is a relatively new concept amongst the agricultural community.

Example of Wild Pollinators Servicing Crops

Papaya (Carica papaya) is a widespread fruit crop throughout the tropics. It is a perennial tree crop that requires adequate pollination in order to set fruit. In tropical and sub-tropical climates, fruit set occurs throughout the year. Sale of fresh papaya provides regular income across the season for many smallholder farmers; for example in Kenya provides single fruits are sold for between 20-100 Kshs ($ 0.26- 1.3), depending on the location and local abundance/availability of fruit. Most smallholder farms produce at least fifty individual saleable fruits a season. The trees are often grown in riparian areas as well as the edges of cultivated fields and as hedgerows within farmsteads. Papaya is pollinated primarily by hawkmoths (Sphingidae).

The species of hawkmoths vary from site to site, but in general, any medium-tolarge- bodied, relatively long-tongued species of hawkmoth can serve as a pollinator. Hawkmoths require adequate areas of habitat for larval food-plants, sheltered diurnal nesting sites, areas for courtship and mating, and other high-energy floral resources in the form of wildflowers. Farmers need to protect and encourage hawkmoths for adequate fruit set on papaya. Farms located within wild areas have high yields and traditionally produce the best-tasting fruit. In addition, as many of these moths travel long distances, they provide benefits to farmers from wild or protected areas adjacent to agricultural landscapes. Yet farmer awareness of the importance of pollination for quality papaya production, and of the need to maintain male trees even though they do not produce fruit, is weak.

Consumer Concerns with Pollination

Any valuation has to be done from the perspective of both consumers and producers of pollination services. While one usually thinks of pollination as being of value to farmers, the consumer perspective should not be left out. An analysis of the economics of pollinator deficits concluded that consumers of a commodity affected by a pollinator deficit may suffer because the commodity costs more and becomes less available. Consumers may thus have to pay more for traded commodities because of pollinator declines.

The foodstuffs for which pollinators are important are largely the fruits and vegetables that provide critical nutrients and minerals, and that in an increasingly industrialised world are disappearing from people's diets. This is a concern, not just to the developed world, but to the developing world, in a number of respects. When any population is surviving on limited caloric intake due to poverty, natural disasters or political instability, the contribution and quantities of pollinator-dependent fruits and vegetables in their diet may be of critical importance to their health. Secondly, despite a conception (from many surveys of consumers) that fruits and vegetables are expensive compared to other foods, the comparative cost per serving of more than

50 common fruits and vegetables has been calculated as far below almost all other, (and many less nutritious) foodstuffs.

As noted above, well-pollinated crops can be of noticeably better quality, and consumers and markets are sensitive to quality considerations: in Canada, good pollination in apple orchards resulted in about one extra seed per apple, which produced larger and more symmetrical apples. These improved apples were estimated to provide marginal returns of about 5-6%, or about Can.

Pollination's Contribution to Seed Production and Plant Genetic Diversity

Many crops, through the selective breeding and replication practices of humans, lose their genetic diversity over time. Exposure to pollinators may be one means of introducing a selective influence to maintain genetic diversity. Studies on bottle-gourd in Kenya have shown how important a diverse pollinator community is to maintaining the extraordinarily diverse forms of gourds.

Plant breeders have traditionally not been concerned with selecting plants for their attractiveness to pollinators. Yet plant genetic make up can affect the level of pollination a crop plant may receive. In many cases, pollinators favour one variety over another, despite the close proximity of different varieties. Farmers may benefit from understanding that strategic plantings, alternating different varieties of chilli peppers in a checkerboard pattern for example, can optimize effective pollination visits to two varieties of different attractiveness, and at the same time promote cross-hybridization and better fruit production. Nectarless cultivars of melon have little attraction for pollinators, and need to be planted in fields with cultivars providing nectar in order to achieve sufficient pollination. Landraces and cultivars that have retained characteristics attractive to pollinators is a little-appreciated aspect of plant genetic diversity that merits conservation.

While good pollination is not a factor in the production of leafy vegetables and root crops, it does have a greatly underappreciated importance in the seed production of such commodities. Estimates of increased seed set due to pollinators have been made in different parts of the world; assured pollination has been variously responsible for increases in seed yield of 22- 100 percent (radish), 100-300 percent (cabbage), 100-125 percent (turnip), 91-135 percent (carrot) and 350-9 000 percent (onion).

Pollination as a Forage/Livestock-related Genetic Resource

Globally, the most important forage crop- alfalfa- relies almost entirely on insect visitors for seed production. Many other sown pasture crops such as clovers that are expected to selfseed, also produce far greater quantities of seed when they receive visits from bees. Grazing systems that depend on livestock obtaining their feed from native vegetation may have a strong dependence on pollinators, for forb and browse species reproduction and tree pods that are eaten by grazing animals. In the Horn of Africa, for example, the short-lived browse plant Indigofera which forms the basis of camel diets in arid rangelands, is pollinated by a least five wild bees. In the same region, Acacia seed pods are resource of high, and sometimes unrecognised potential, contributing directly or indirectly to pastoral communities' livelihoods and survival. Their main use is for livestock fodder, but they also are bought and sold and form a food of last resort in times of drought. The pods depend on visits of a diversity of visitors- bees, ants, wasps, butterflies, moths, sunbirds and beetle to Acacia flowers.

Successful rangeland revegetation may be enhanced by strategies that consider the role of pollination, seed-dispersal and other plant-animal interactions in ecosystem health and recovery. Many perennial forbs and shrubs require animals for successful pollination, reproduction, and subsequent maintenance of species on a site; however, pollination biology of many rangeland plants and pollinator abundances at potential revegetation sites are largely unknown. Large-area aerial insect control programmes, such as desert locust control in Africa or rangeland grasshopper control in North America may have impacts on non-target species such as bees that need to forage over wide areas in arid ecosystems, and such impacts should be minimized.

Pollination as a Tool in Adaptation to Changing Environments and Minimising Risk

Climate change is causing changes in the distribution of many species. There is an interest in identifying crop genetic resources that help crops adapt to climate change. Pollinators, however, will largely respond by contracting or expanding their ranges according to new climatic patterns. Thus the possibility of crops losing key pollinating species, or mismatches in the ranges of crops and their pollinators, is a real threat.

Such effects have already been felt in the seed industry of India.. Since seed production requires a certain degree of chilling to induce seed formation in temperate crops, many vegetable seed farms are located in mountainous regions, such as the Hindu-Kush Himalayas. While mountainous regions can provide such a climate, they also make farmers increasingly vulnerable to the effects of climate change. Farmers in the Kullu valley of Himachal Pradesh state in India are finding that overall temperatures have been rising, while rains have become more unpredictable, leading to several crop failures. Vegetable seed yields have been decreasing, yet the challenge of ensuring sufficient natural pollination under changing climatic conditions has not been addressed by researchers, much less farming communities.

Resilience is built in agroecosystems through biodiversity. Crops produce optimally with a suite of pollinators possibly including, but not limited to managed honeybees. Different pollinators become most active at different times of the day or under different weather conditions, and even between years the most abundant and effective pollinators of a crop may shift from one pollinator to another. A diverse assemblage of pollinators, with different traits and responses to ambient conditions, is one of the best ways of minimizing risks due to climatic change. The "insurance" provided by a diversity of pollinators ensures that there are effective pollinators not just for current conditions, but for future conditions as well. A biodiverse agroecosystem, with many more facilitative interactions between crops and crop-associated biodiversity, may also contribute significantly to carbon sequestration.

Pollination may also help to counteract the tendency of crops, through the selective breeding and replication practices of humans, to lose their genetic diversity over time. Exposure to pollinators may be one means of introducing a selective influence to maintain genetic diversity. For example, the vast fields of blue agave grown for tequila production that cover the western state of Jalisco in Mexico have been hard-hit by disease. The disease could cause such decimation in part because, through a long and sophisticated process of artificial selection, virtually all the blue agave grown in this region are actually clones of only two plants. It has been suggested that the disease impact on these agaves was particularly severe due to the low genetic diversity. It has been proposed to manage a small portion of the blue agave plantations in a way that will permit exchange of pollen (through bat pollinators) with wild agave types in the nearby barrancas, so that selection of genetic material with resistance to various environmental stresses is conserved.

Threats and Risks to Pollination Services

The value of pollinators for sustainable livelihoods has thus been detailed, for crop production, for seed production and maintenance of crop genetic diversity, for forage resources and for adaptation to climate change and environmental stress. Smallholder and subsistence agricultural systems have traditionally been based on practices that promote high diversity onfarm. Modern pressures that are changing farming practices and driving overall environmental change present the following risks to loss of pollinators' services:

Habitats required by many pollinators are being lost through changing land-use patterns such as increasing agricultural intensification. Pollinators require a range of resources from their environment for foraging, nesting, reproduction and shelter. The loss of any one of these requirements can cause pollinators to become locally extinct.

Excessive use or inappropriate application of pesticides and other agro-chemicals is known to have negative impacts on a range of pollinators.

Climate change may potentially be one of the most severe threats to pollinator biodiversity. Substantial distribution changes are predicted for groups such as butterflies31.

Invasive species are globally recognised to have major negative impacts across a wide range of taxa. Two major causes of honeybee declines globally are parasitic mites (Varroa jacobsoni and Acarapsis woodi) and the expansion of the range of Africanized honeybees in the US.

The basis for a more sustainable path of agricultural growth lies in maintaining functional biodiversity in farming systems. The deliberate conservation of pollinators- and its synergy with integrated pest control- offers ways to maintain yields while reducing purchased inputs. Many of the measures that promote pollinators can also promote other ecosystem services such as soil improvement by cover cropping, increasing the abundance of diverse soil functional groups; habitat management of natural enemies for pest management; breaking cycles of damaging pests through greater crop diversity, or erosion control through contour plantings and hedgerows. The viability of rural lifestyles that engage in such practices of maintaining high diversity on farms can be recognised and supported by an enabling policy environment. Conversely the pressures of rapidly commercialising agriculture, such as with the horticultural sector in Africa, can lead to the adoption of practices (intensification, greater use of agrochemicals, larger field sizes) that negatively impact pollination services, if applied in the absence of deliberate efforts to conserve and sustain them.

International Response

Over the decade, the international community has increasingly recognised the importance of pollinators as an element of agricultural diversity supporting human livelihoods. Yet mounting evidence points to a potentially serious decline in populations of pollinators. In response to this, the Conference of the Parties to the Convention on Biological Diversity has made conservation and sustainable use of pollinators a priority, establishing an International Initiative for the Conservation and Sustainable Use of Pollinators (IPI) and requested FAO to facilitate and coordinate the Initiative. It also requested the preparation of a plan of action. The Plan of Action of the IPI, as prepared by FAO and the CBD secretariat and adopted at COP 6 (decision VI/5), is structured around four elements: assessment, adaptive management, capacity building and mainstreaming. The Plan of Action of the IPI presents a cohesive set of actions that address barriers serve to create awareness and build capacity to conserve and sustainably use pollination services.

FAO has addressed its coordination role within the International Pollinator Initiative, through leveraging limited resources to, inter alia,

i. Facilitate the development of the African Pollinator Initiative, resulting in the formation of a network, the publication of its Plan of Action, a special issue of a journal dedicated to pollination work in Africa, and an initial stocktaking report of the importance of pollinators in crop and browse production in Africa.

ii. Facilitate a series of case studies from Pakistan, India, China and Nepal on the role of pollination in contributing to human livelihoods.

iii. Support to two international conferences in Brazil to refine methods of monitoring and assessment of pollination services.

iv. Undertake the development of a global project to harness the benefits of pollination services provided by wild biodiversity for human livelihoods and sustainable agriculture, through an ecosystem approach, with on the ground activities in Brazil, Ghana, India, Kenya, Nepal, Pakistan and South Africa.

v. Initiate pilot projects in Ghana, India, Laos, Mali and Nepal to build capacity for communities and institutions to develop pollination management plans

vi. Serve as a source of information on crop pollination services, through the compilation of case studies, and relevant databases.

FAO also undertakes a number of other activities that address measures important to pollination services, such as integrated pest management, and conservation agriculture.

A further articulation of an ecosystem approach to the management and sustainable use of crop-related genetic resources such as pollinators and other aspects of crop-associated biodiversity, would include:

An assessment of status and trends of agricultural ecosystem services;

Development of guidelines for best practices to sustain these services;

Production of guidelines on applying adaptive management to mitigate negative impacts of new technologies on crop-associated biodiversity, and

Production of guidelines on incorporation of measures to sustain ecosystem services of importance to agriculture into policy instruments at different levels, including local land use planning, national biodiversity strategies and action plans, and the planning process in the agricultural sector.

The Commission on Genetic Resources for Food and Agriculture also has recognised that the relevant genetic resources include not just crops and livestock, but the full suite of croprelated biodiversity that serves to maintain productivity, and sustain agroecosystem functions. In 1995, the Commission broadened its mandate (Resolution 3/95), to cover all components of biodiversity of relevance to food and agriculture, specifically noting that countries need an integrated approach to deal with policy issues regarding biological diversity.

Bibliography

Behre, K-E: *Anthropogenic Indicators in Pollen Diagrams*, A.A. Balkema, Rotterdam, 1986.

Bertenschlager, S Auinger, A. Simonsberger, P.: *Pollen Morphology of the Achatocarpaceae*. Ber. Nat-med. Ver. Innsbruck., 1972.

Erdtman, G.: *Pollen Morphology and Plant Taxonomy, Angiosperms*, Printed by Almquist Wiksell, Stockholm, Sweden, 1952.

Faegri, K. and J. Iversen: *Textbook of Pollen Analysis*, John Wiley & Sons, New York, 1989.

Fuhsiung, W, Nanfen, C., Yulong, Z, and Huiqiu, Y.: *Pollen Flora of China*, Academia Sinica, 1995.

Gonzalez Quintero, Lauro: *Flora Polinica y tipos de vegetacion del Valle del Mezquital*, National School of Biological Sciences, Mexico City, Mexico, 1967.

Heusser, C.J.: *Pollen and Spores of Chile*, University of Arizona Press, Tucson, 1971.

Horowitz, A.: *Palynology of Arid Lands*, Elsevier, Amsterdam, 1992.

Ikuse, M.: *Pollen Grains of Japan*, Hirokawa Publishing Co., Tokyo, 1956.

Lewis, W.H., Prathibha, V., and Zenger, V.E.: *Airborne and Allergenic Pollen of North America*, Johns Hopkins Univ. Press, Baltimore, 1983.

Markgraf, V. and H. L. D'Antoni: *Pollen Flora of Argentina*, University of Arizona Press, 1978.

McAndrews, J.H., A.A. Berti, and G. Norris: *Key to the Pollen and Spores of the Great Lakes Region*, Life Science Miscellaneous Publication, Royal Ontario Museum, 1973.

Moore, P.D., J.A. Webb, and Collinson, M.E.: *Pollen Analysis*, Blackwell Scientific, London, 1991.

Nari, P.K.K.: *Pollen Grains of Indian Plants - III*, Natioinal Botanic Gardens, Lucknow, India, 1962.

Ogden , E.C., Raynor, G.S., Hayes, J.V., Lewis, D.M., and Haines, J.H.: *Manual for Sampling Airborne Pollen*, Hafner Press, N.Y., 1974.

Tschudy, R.H. and Scott, R.A.: *Aspects of Palynology*, John Wiley & Sons, New York, 1969.

INDEX

A

Acid Rain, 270.
Agar Production, 50, 101, 106, 107, 110, 113, 123, 138.
Agarophytes, 48, 52, 53, 54, 59, 67, 76, 82, 93, 94, 95, 105, 106, 107, 110, 117, 124, 125, 138, 139, 147, 148, 149.
Agency, 2, 12, 47, 120.
Agriculture, 7, 10, 153, 174, 182, 226, 275, 286, 287, 288.
Agroforestry Systems, 264.
Air Pollution, 18.
Applications, 3, 14, 17, 20, 33, 54, 107, 250.
Approaches, 6, 13, 20, 242, 243, 268.
Assumptions, 113, 114, 162.
Atmosphere, 4.

B

Biodiversity, 5, 6, 7, 220, 221, 225, 226, 231, 232, 233, 236, 240, 242, 249, 250, 253, 259, 260, 261, 264, 266, 267, 268, 269, 270, 272, 273, 275, 276, 284, 285, 286, 287, 288.
Biological Diversity, 5, 6, 7, 273, 286.

C

Capsicum Species, 149, 150, 152, 165, 166, 185, 186, 207.
Carbon Dioxide, 4, 14.
Collections, 28, 170, 188, 203, 262.
Commission, 14, 18, 119, 288.
Committee, 1.
Community, 4, 5, 9, 12, 15, 22, 28, 34, 42, 103, 136, 222, 230, 232, 234, 239, 241, 247, 250, 257, 260, 262, 275, 279, 280, 282, 286.
Conditions, 25, 28, 45, 46, 50, 51, 55, 56, 59, 60, 63, 68, 69, 70, 71, 73, 78, 79, 82, 91, 94, 96, 97, 106, 114, 117, 118, 121, 139, 150, 156, 157, 158, 159, 170,

186, 188, 189, 193, 202, 204, 205, 207, 213, 214, 215, 220, 225, 226, 227, 229, 233, 250, 252, 263, 277, 279, 285.
Conference, 10, 13, 286.
Conservation, 16, 48, 149, 154, 187, 188, 198, 221, 223, 224, 225, 226, 229, 232, 233, 234, 235, 236, 239, 241, 242, 243, 246, 249, 260, 261, 262, 264, 265, 266, 267, 268, 269, 270, 272, 273, 274, 275, 280, 282, 286, 287, 288.
Considerations, 55, 131, 134, 139, 143, 145, 279, 282.
Consumption, 50, 98, 99, 120, 125, 128, 131, 153, 245.
Convention, 5, 6, 7, 28, 286.
Correlation, 91, 168, 169, 170, 180, 181, 182, 183, 184.
Cryopreservation, 185, 186, 196.
Culture, 44, 47, 48, 49, 54, 55, 56, 57, 59, 60, 61, 62, 64, 65, 67, 68, 72, 73, 74, 75, 76, 77, 78, 79, 80, 83, 84, 85, 86, 87, 88, 89, 91, 92, 106, 107, 109, 113, 114, 124, 125, 133, 135, 138, 148, 189, 190, 193, 201, 214, 215, 216, 269.

D

Data Access, 22, 30, 36.
Data Processing Bundle, 29.
Development, 7, 8, 10, 11, 18, 19, 29, 31, 33, 34, 47, 48, 54, 64, 68, 75, 80, 86, 87, 89, 100, 107, 117, 118, 123, 125, 133, 134, 140, 143, 147, 148, 149, 159, 164, 190, 191, 192, 193, 195, 204, 207, 209, 214, 215, 216, 217, 220, 221, 222, 242, 252, 261, 262, 263, 265, 270, 272, 277, 278, 287, 288.
Dicotyledons, 231.
Distribution, 2, 5, 7, 9, 17, 19, 23, 29, 31, 35, 36, 37, 42, 58, 77, 78, 88, 116, 127, 130, 141, 146, 164, 221, 224, 225, 227, 231, 232, 234, 236, 241, 253, 256, 284, 286.

E

Earth Observation Satellites, 1.
Ecosystem, 6, 14, 33, 222, 223, 224, 226, 232, 236, 241, 244, 246, 249, 253, 260, 269, 272, 275, 276, 279, 280, 283, 286, 287, 288.

Emissions, 7, 14, 27, 270.
Environment, 9, 14, 35, 37, 44, 156, 168, 223, 225, 257, 258, 265, 266, 269, 285, 286.
Evaluation, 34, 36, 41, 55, 66, 94, 95, 112, 135, 151, 154, 155, 167, 178, 180, 199, 200, 201, 206, 221, 222, 223, 224, 232, 271.
Exploitation, 82, 120, 139, 140, 162, 178, 214, 260.
Exports, 100, 106, 118, 136, 137, 148, 277.
Extraction, 6, 50, 94, 95, 96, 97, 98, 99, 100, 104, 105, 106, 107, 108, 110, 111, 112, 113, 114, 150, 173, 255, 274.

F

Forest Ecosystems, 19, 247, 253, 269.
Forest Management, 5, 6, 11, 13, 39, 260, 269.
Fragmentation, 3, 5, 7, 220, 226, 229, 238, 239.
Fucus Vesiculosus, 42, 44, 45, 46.

G

Germination, 68, 69, 70, 73, 186, 187, 196, 197, 204, 207, 208, 209, 210, 247.
Global Observations, 1, 2, 3.
Gracilaria, 46, 47, 48, 49, 50, 51, 52, 53, 55, 56, 57, 58, 59, 60, 61, 62, 63, 64, 65, 66, 67, 68, 69, 70, 71, 72, 73, 74, 75, 76, 77, 78, 80, 81, 82, 83, 84, 85, 86, 87, 88, 89, 90, 91, 93, 94, 95, 96, 97, 98, 99, 101, 102, 104, 105, 106, 107, 110, 111, 112, 116, 117, 118, 120, 121, 122, 124, 125, 133, 138, 139, 140, 142, 143, 145, 146, 147, 148.

H

Heritability, 164, 168, 169, 173, 174.
Horticultural Crops, 275, 277.
Hybridization, 59, 177, 205, 213, 282.
Hydrological Research, 250.
Hydrolysis, 94, 95, 97.

I

Imports, 105, 106, 107, 121, 122, 126, 130, 131, 134, 135, 136, 137.
Improvements, 5, 21, 277.
Industry, 47, 48, 51, 54, 75, 95, 99, 100, 105, 106, 110, 111, 117, 118, 120, 121, 122, 123, 124, 133, 134, 148, 167, 174, 284.

Information, 1, 4, 6, 7, 8, 9, 10, 11, 18, 19, 20, 22, 23, 25, 26, 27, 29, 30, 31, 32, 33, 34, 35, 37, 38, 39, 47, 51, 77, 100, 121, 129, 132, 135, 139, 140, 141, 142, 143, 144, 148, 153, 155, 156, 170, 175, 216, 223, 225, 234, 236, 240, 243, 249, 265, 275, 280, 288.

Instruments, 7, 32, 288.

Invertebrates, 49, 233, 238, 249.

M

Maintenance, 31, 61, 64, 76, 87, 154, 221, 222, 225, 264, 276, 283, 285.

Management, 2, 4, 5, 6, 10, 11, 13, 19, 20, 22, 23, 26, 27, 31, 36, 39, 48, 52, 53, 54, 61, 62, 64, 74, 127, 142, 143, 144, 146, 147, 148, 150, 153, 205, 225, 227, 229, 231, 235, 237, 238, 239, 241, 244, 247, 253, 255, 257, 258, 260, 261, 262, 264, 265, 270, 271, 273, 275, 276, 280, 282, 284, 288.

Manufacture, 54, 99, 107, 122, 125, 129, 138, 139, 148.

Marketing, 47, 52, 87, 100, 107, 117, 119, 121, 122, 123, 124, 127, 128, 129, 130, 132, 133, 134.

Measurements, 27, 28, 85, 113, 143, 161, 252, 263.

Mechanism, 23, 31, 35, 39, 58, 193.

Methylation, 207, 208, 209, 210.

Mountainous, 23, 250, 276, 284.

N

Natural Resources, 47, 48, 113, 138, 249, 257, 270.

Nutrition, 85, 216, 219, 275, 277.

O

Observations, 1, 2, 3, 7, 8, 14, 23, 27, 28, 33, 43, 111, 171, 177, 179, 182, 183, 186, 187, 191, 213, 243.

P

Pesticides, 220, 222, 286.

Phenomena, 207.

Pilot Project, 2, 3, 11, 12, 24, 28, 29, 31, 33, 35, 36, 38, 39, 47, 87, 125.

Policy, 4, 6, 7, 8, 13, 18, 19, 22, 35, 39, 51, 123, 249, 254, 257,

259, 260, 261, 262, 269, 271, 286, 288.
Pollen Collection, 186, 187.
Pollen Cryopreservation, 185.
Pollen Viability, 160, 162, 187.
Pollination, 160, 161, 162, 192, 213, 235, 244, 245, 246, 247, 248, 276, 277, 278, 279, 280, 281, 282, 283, 284, 285, 286, 287, 288.
Pollinators, 213, 245, 247, 248, 275, 276, 277, 278, 279, 280, 281, 282, 283, 284, 285, 286, 287, 288.
Pollution, 10, 18, 42, 253, 270.
Pond Management, 61, 62.
Prevention, 5.
Product Distribution, 23, 35.
Product Validation, 23, 33, 34, 35.
Production, 4, 5, 10, 11, 18, 24, 32, 46, 47, 48, 49, 50, 51, 52, 54, 56, 57, 58, 59, 60, 64, 66, 67, 68, 74, 75, 78, 82, 84, 86, 87, 88, 89, 92, 96, 100, 101, 104, 105, 106, 107, 108, 109, 110, 112, 113, 114, 115, 117, 121, 122, 123, 125, 133, 135, 137, 138, 139, 140, 141, 142, 143, 144, 145, 146, 148, 149, 156, 161, 162, 185, 188, 190, 202, 213, 214, 216, 248, 263, 270, 275, 277, 279, 280, 281, 282, 283, 284, 285, 287, 288.
Products, 2, 3, 4, 6, 9, 11, 12, 16, 21, 23, 24, 25, 26, 27, 28, 29, 33, 34, 35, 37, 38, 40, 41, 52, 54, 77, 78, 89, 94, 96, 99, 100, 121, 122, 123, 126, 129, 147, 148, 150, 164, 167, 262, 263.
Project, 2, 3, 4, 9, 11, 12, 13, 14, 15, 16, 17, 18, 19, 20, 21, 22, 23, 24, 25, 26, 28, 29, 30, 31, 32, 33, 34, 35, 36, 37, 38, 39, 40, 41, 47, 87, 110, 118, 119, 120, 121, 131, 159, 181, 227, 248, 259, 267, 284, 288.
Protection, 5, 11, 48, 144, 145, 230, 233, 256, 260, 261, 262, 270.
Protocol, 7, 150, 215, 242, 243.

R

Rainforest, 16, 22, 226, 231, 238, 239, 247, 260, 273.
Relationships, 113, 115, 149, 150, 152, 249, 277.
Research, 1, 2, 3, 4, 8, 9, 10, 14, 15, 17, 18, 20,

21, 28, 32, 38, 43, 46, 47, 48, 51, 77, 78, 89, 90, 107, 110, 111, 112, 123, 124, 125, 126, 130, 135, 146, 153, 154, 157, 166, 167, 181, 188, 190, 199, 210, 220, 236, 238, 242, 247, 249, 250, 254, 255, 256, 267, 268, 271, 272, 273.

Resistance, 90, 150, 153, 154, 159, 164, 178, 179, 180, 185, 187, 198, 199, 200, 201, 202, 203, 204, 205, 206, 207, 244, 285.

Resource Management, 4, 22, 53.

Restoration, 221, 223, 224, 225, 226, 230, 231, 232, 241, 271, 280.

S

Satellites, 1, 3, 14, 27, 28.

Scelionidae, 238.

Strategy, 1, 18, 24, 25, 29, 32, 34, 88, 166, 185, 190, 202, 233.

T

Technology, 10, 11, 13, 19, 20, 21, 22, 48, 51, 78, 86, 87, 88, 99, 107, 108, 110, 111, 133, 157, 181, 210, 225, 243, 271, 272.

Temperature, 4, 33, 44, 45, 46, 59, 60, 62, 64, 69, 70, 71, 78, 80, 82, 83, 84, 85, 93, 96, 97, 98, 100, 108, 112, 159, 160, 161, 162, 187, 193, 197, 198, 203, 206, 213, 237.

Transportation, 46, 94, 95, 119, 132.

Treatment, 46, 51, 73, 94, 95, 96, 97, 98, 102, 104, 106, 162, 181, 189, 197, 202, 211, 215, 216.

Tropical Agarophytes, 54, 138.

U

Utilization, 10, 46, 59, 75, 121, 139, 228, 268.

V

Validation, 8, 11, 21, 23, 25, 30, 33, 34, 35, 41.

Variation, 3, 66, 91, 142, 143, 144, 148, 156, 158, 167, 168, 169, 173, 174, 180, 187, 212, 246, 248.

Vegetation, 14, 18, 19, 27, 30, 193, 227, 241, 245, 257, 271, 283.

W

World Agar Production, 50.

□□□